PENGUIN BOOKS

HOW TO AVOID A CLIMATE DISASTER

'The most comprehensible explanation for what's driving our warming
planet; how to measure the impact of the myriad contributions to this
staggering and seemingly incalculable problem; and ultimately how to go
about finding more effective approaches to each of them. It's the closest
thing I've seen to a how-to guide for addressing the climate crisis' *Fortune*

'Clear, concise on a colossal subject, and intelligently holistic in its approach
to the problem . . . A fine primer on how to get ourselves out of this mess'
New Scientist

'An optimistic account of how climate change might be solved without
destroying the world's economies in the process' *The Times*, Books of the Year

'Compelling . . . Gates is a serious and genuine force for
good on climate change' *Observer*

'Gates gathers advice from experts while laying out his vision for
technological innovations that could reduce greenhouse gases and stop
the warming of the planet. If even some of his plans work, this might be
the most important book of the year' *CNN*

'Practical and accessible' *USA Today*

'Concise, straightforward . . . Gates has crafted a calm, reasoned,
well-sourced explanation of the greatest challenge of our time and what
we must change to avoid cooking our planet' *Associated Press*

'His expertise . . . is apparent in the book's lucid explanations of
the scientific aspects of climate change. The solutions he outlines
are pragmatic and grounded in forward-thinking economic reasoning.
Although he does not avoid the hard truths we must face as our
climate changes, Gates remains optimistic and believes that we have
the ability to avoid a total climate disaster' *Science*

'The author's enthusiasm and curiosity about the way things work is infectious. He walks us through not just the basic science of global warming, but all the ways that our modern lives contribute to it . . . Gates seems energized by the sheer size and complexity of the challenge. That's one of the best things about the book – the can-do optimism and conviction that science in partnership with industry are up to the task' *The Christian Science Monitor*

'One of the most accessible, practical, and interesting books on the topic to emerge since Al Gore's *An Inconvenient Truth*' *Oprah Daily*

'Gates's carefully packaged nuggets of information are not only easy to understand, but they aim to provide the reader with practical tools to engage with the density of climate change information . . . What Gates has achieved with his book is something rare . . . A solutions-based strategy that is as informed on the commercial realities of scaling new technologies as it is on the environmental consequences of not doing so' *Business Post*

ABOUT THE AUTHOR

Bill Gates is a technologist, business leader, and philanthropist. In 1975, he cofounded Microsoft with his childhood friend Paul Allen; today he is cochair of the Bill & Melinda Gates Foundation. He also launched Breakthrough Energy, an effort to commercialize clean energy and other climate-related technologies.

HOW TO AVOID A CLIMATE DISASTER

The Solutions We Have and the Breakthroughs We Need

BILL GATES

PENGUIN BOOKS

PENGUIN BOOKS

UK | USA | Canada | Ireland | Australia
India | New Zealand | South Africa

Penguin Books is part of the Penguin Random House group of companies
whose addresses can be found at global.penguinrandomhouse.com.

This edition published by arrangement with Alfred A. Knopf,
a division of Penguin Random House LLC 2021
First published in Great Britain by Allen Lane 2021
Published in Penguin Books 2022
002

Printed and bound in Great Britain by Clays Ltd, Elcograf S.p.A.

The authorized representative in the EEA is Penguin Random House Ireland,
Morrison Chambers, 32 Nassau Street, Dublin D02 YH68

A CIP catalogue record for this book is available from the British Library

ISBN: 978–0–141–99301–0

www.greenpenguin.co.uk

To the scientists, innovators, and activists
who are leading the way

CONTENTS

HOW TO AVOID A CLIMATE DISASTER

52 BILLION TO ZERO

There are two numbers you need to know about climate change. The first is 52 billion. The other is zero.

Fifty-two billion is how many tons of greenhouse gases the world typically adds to the atmosphere every year. Although the figure may go up or down a bit from year to year, it's generally increasing. This is *where we are today.**

Zero is *what we need to aim for*. To stop the warming and avoid the worst effects of climate change—and these effects will be very bad—humans need to stop adding greenhouse gases to the atmosphere.

This sounds difficult, because it will be. The world has never done anything quite this big. Every country will need to change its ways. Virtually every activity in modern life—growing things, making things, getting around from place to place—involves releasing greenhouse gases, and as time goes on, more people will be living this modern lifestyle. That's good, because it means their lives are

---

* Fifty-two billion tons is based on the latest available data. Global emissions dropped a bit in 2020—around 4.5 percent—because the COVID-19 pandemic slowed the economy so dramatically. But then emissions likely rose about 6 percent in 2021 as the economy recovered. So I will use 52 billion tons as the total. We'll return to the subject of COVID-19 periodically throughout this book.

getting better. Yet if nothing else changes, the world will keep producing greenhouse gases, climate change will keep getting worse, and the impact on humans will in all likelihood be catastrophic.

But "if nothing else changes" is a big If. I believe that things *can* change. We already have some of the tools we need, and as for those we don't yet have, everything I've learned about climate and technology makes me optimistic that we can invent them, deploy them, and, if we act fast enough, avoid a climate catastrophe.

This book is about what it will take and why I think we can do it.

Two decades ago, I would never have predicted that one day I would be talking in public about climate change, much less writing a book about it. My background is in software, not climate science, and these days my full-time job is working with my wife, Melinda, at the Gates Foundation, where we are super-focused on global health, development, and U.S. education.

I came to focus on climate change in an indirect way—through the problem of energy poverty.

In the early 2000s, when our foundation was just starting out, I began traveling to low-income countries in sub-Saharan Africa and South Asia so I could learn more about child mortality, HIV, and the other big problems we were working on. But my mind was not always on diseases. I would fly into major cities, look out the window, and think, *Why is it so dark out there? Where are all the lights I'd see if this were New York, Paris, or Beijing?*

In Lagos, Nigeria, I traveled down unlit streets where people were huddling around fires they had built in old oil barrels. In remote villages, Melinda and I met women and girls who spent hours every day collecting firewood so they could cook over an open flame in their homes. We met kids who did their homework by candlelight because their homes didn't have electricity.

I learned that about a billion people didn't have reliable access to

Melinda and I often meet children like nine-year-old Ovulube Chinachi, who lives in Lagos, Nigeria, and does his homework by candlelight.

electricity and that half of them lived in sub-Saharan Africa. (The picture has improved a bit since then; today roughly 860 million people don't have electricity.) I thought about our foundation's motto—"Everyone deserves the chance to live a healthy and productive life"—and how it's hard to stay healthy if your local medical clinic can't keep vaccines cold because the refrigerators don't work. It's hard to be productive if you don't have lights to read by. And it's impossible to build an economy where everyone has job opportunities if you don't have massive amounts of reliable, affordable electricity for offices, factories, and call centers.

Around the same time, the late scientist David MacKay, a professor at Cambridge University, shared a graph with me that showed the relationship between income and energy use—a country's per capita income and the amount of electricity used by its people. The chart plotted various countries' per capita income on one axis and energy consumption on the other—and made it abundantly clear to me that the two go together:

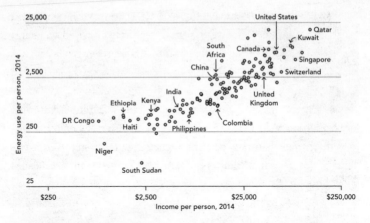

Income and energy use go hand in hand. David MacKay showed me a chart like this one plotting energy consumption and income per person. The connection is unmistakable. (IEA; World Bank)

As all this information sank in, I began to think about how the world could make energy affordable and reliable for the poor. It didn't make sense for our foundation to take on this huge problem—we needed it to stay focused on its core mission—but I started kicking around ideas with some inventor friends of mine. I read more deeply on the subject, including several eye-opening books by the scientist and historian Vaclav Smil, who helped me understand just how critical energy is to modern civilization.

At the time, I didn't understand that we needed to get to zero. The rich countries that are responsible for most emissions were starting to pay attention to climate change, and I thought that would be enough. My contribution, I believed, would be to advocate for making reliable energy affordable for the poor.

For one thing, they have the most to gain from it. Cheaper energy would mean not only lights at night but also cheaper fertilizer for their fields and cement for their homes. And when it comes to climate change, the poor have the most to lose. The majority of them are farmers who already live on the edge and can't withstand more droughts and floods.

Things changed for me in late 2006 when I met with two former Microsoft colleagues who were starting nonprofits focused on energy and climate. They brought along two climate scientists who were well versed in the issues, and the four of them showed me the data connecting greenhouse gas emissions to climate change.

I knew that greenhouse gases were making the temperature rise, but I had assumed that there were cyclical variations or other factors that would naturally prevent a true climate disaster. And it was hard to accept that as long as humans kept emitting any amount of greenhouse gases, temperatures would keep going up.

I went back to the group several times with follow-up questions. Eventually it sank in. The world needs to provide more energy so the poorest can thrive, but we need to provide that energy without releasing any more greenhouse gases.

Now the problem seemed even harder. It wasn't enough to deliver cheap, reliable energy for the poor. It also had to be clean.

I kept learning everything I could about climate change. I met with experts on climate and energy, agriculture, oceans, sea levels, glaciers, power lines, and more. I read the reports issued by the Intergovernmental Panel on Climate Change (IPCC), the UN panel that establishes the scientific consensus on this subject. I watched *Earth's Changing Climate,* a series of fantastic video lectures by Professor Richard Wolfson available through the Great Courses series. I read *Weather for Dummies,* still one of the best books on weather that I've found.

One thing that became clear to me was that our current sources of renewable energy—wind and solar, mostly—could make a big dent in the problem, but we weren't doing enough to deploy them.*

---

* Hydropower—electricity created by water pouring through a dam—is another renewable source, in fact the biggest source of renewable energy in the United States. But we've already tapped most of the available hydropower. There's not a lot of room to grow there. Most of the additional clean energy we want will have to come from another source.

It also became clear why, on their own, they aren't enough to get us all the way to zero. The wind doesn't always blow and the sun doesn't always shine, and we don't have affordable batteries that can store city-sized amounts of energy for long enough. Besides, making electricity accounts for only 26 percent of all greenhouse gas emissions. Even if we had a huge breakthrough in batteries, we would still need to get rid of the other 74 percent.

Within a few years, I had become convinced of three things:

1. To avoid a climate disaster, we have to get to zero.
2. We need to deploy the tools we already have, like solar and wind, faster and smarter.
3. And we need to create and roll out breakthrough technologies that can take us the rest of the way.

The case for zero was, and is, rock solid. Unless we stop adding greenhouse gases to the atmosphere, the temperature will keep going up. Here's an analogy that's especially helpful: The climate is like a bathtub that's slowly filling up with water. Even if we slow the flow of water to a trickle, the tub will eventually fill up and water will come spilling out onto the floor. That's the disaster we have to prevent. Setting a goal to only reduce our emissions—but not eliminate them—won't do it. The only sensible goal is zero. (For more on zero, what I mean by it, and the impact of climate change, see chapter 1.)

But at the time I learned all this, I wasn't looking for another issue to take on. Melinda and I had picked global health and development and U.S. education as the two areas where we would learn a great deal, hire teams of experts, and spend our resources. I also saw that many well-known people were putting climate change on the agenda.

So although I got more involved, I didn't make it a top priority. When I could, I read and met with experts. I invested in some clean energy companies, and I put several hundred million dollars into starting a company to design a next-generation nuclear plant that would generate clean electricity and very little nuclear waste. I gave a TED talk called "Innovating to Zero!" But mostly, I kept my attention on the Gates Foundation's work.

Then, in the spring of 2015, I decided that I needed to do more and speak out more. I had been seeing news reports about college students around the United States who were holding sit-ins to demand that their schools' endowments divest from fossil fuels. As part of that movement, the British newspaper *The Guardian* launched a campaign calling on our foundation to sell off the small fraction of its endowment that was invested in fossil-fuel companies. They made a video featuring people from around the world asking me to divest.

I understood why *The Guardian* had singled out our foundation and me. I also admired the activists' passion; I had seen students protesting the Vietnam War, and later the apartheid regime in South Africa, and I knew they had made a real difference. It was inspiring to see this kind of energy directed at climate change.

On the other hand, I kept thinking about what I had witnessed in my travels. India, for example, has a population of 1.4 billion people, many of them among the poorest in the world. I didn't think it was fair for anyone to tell Indians that their children couldn't have lights to study by, or that thousands of Indians should die in heat waves because installing air conditioners is bad for the environment. The only solution I could imagine was to make clean energy so cheap that every country would choose it over fossil fuels.

As much as I appreciated the protesters' passion, I didn't see how divesting alone would stop climate change or help people in poor countries. It was one thing to divest from companies to fight apartheid, a political institution that would (and did) respond to economic pressure. It's another thing to transform the world's energy system—an industry worth roughly $5 trillion a year and the basis for the modern economy—just by selling the stocks of fossil-fuel companies.

I still feel this way today. But I have come to realize that there are other reasons for me not to own the stocks of fossil-fuel companies—namely, I don't want to profit if their stock prices go up because we don't develop zero-carbon alternatives. I'd feel bad if I benefited from a delay in getting to zero. So in 2019, I divested all my direct holdings in oil and gas companies, as did the trust that manages the Gates Foundation's endowment. (I hadn't had money in coal companies in several years.)

This is a personal choice, one that I'm fortunate to be able to make. But I'm well aware that it won't have a real impact on lowering emissions. Getting to zero requires a much broader approach: driving wholesale change using all the tools at our disposal, including government policies, current technology, new inventions, and the ability of private markets to deliver products to huge numbers of people.

Later in 2015 came an opportunity to make the case for innovation and new investments: the COP 21, a major climate change conference to be held by the United Nations in Paris that November and December. A few months before the conference, I met with François Hollande, who was the president of France at the time. Hollande was interested in getting private investors to join the conference, and I was interested in getting innovation on the agenda. We both saw an opportunity. He thought I could help bring investors to the table; I said that made sense, though it would be easier to do if governments also committed to spending more on energy research.

That was not necessarily going to be an easy sell. Even America's investment in energy research was (and still is) far lower than in other essential areas, like health and defense. Although some countries were modestly expanding their research efforts, the levels were still very low. And they were reluctant to do much more unless they knew that there would be enough money from the private sector to take their ideas out of the lab and turn them into products that actually helped their people.

But by 2015, private funding was drying up. Many of the venture capital firms that had invested in green tech were pulling out of the industry because the returns were so low. They were used to investing in biotechnology and information technology, where success often comes quickly and there are fewer government regulations to deal with. Clean energy was a whole other ball game, and they were getting out.

Clearly, we needed to bring in new money and a different approach that was tailored specifically to clean energy. In September, two months before the Paris conference started, I emailed two dozen wealthy acquaintances, hoping to persuade them to commit venture funding to complement the governments' new money for research. Their investments would need to be long term—energy breakthroughs can take decades to develop—and they would have to tolerate a lot of risk. To avoid the potholes that the venture capitalists had run into, I committed to help build a focused team of experts who would vet the companies and help them navigate the complexities of the energy industry.

I was delighted by the response. The first investor said yes in less than four hours. By the time the Paris conference kicked off two months later, 26 more had joined, and we had named it the Breakthrough Energy Coalition. Today, the organization now known as Breakthrough Energy includes philanthropic programs, advocacy efforts, and private funds that have invested in more than 85 companies with promising ideas.

The governments came through too. Twenty heads of state got together in Paris and committed to doubling their funding for research. President Hollande, U.S. President Barack Obama, and Indian Prime Minister Narendra Modi had been instrumental in pulling it together; in fact, Prime Minister Modi came up with the name: Mission Innovation. Today Mission Innovation includes 24 countries and the European Commission and has unlocked $4.6 billion a year in new money for clean energy research, an increase of more than 50 percent in just a handful of years.

Launching Mission Innovation with world leaders at the 2015 UN climate conference in Paris. (See page 237 for the names of those photographed.)

The next turning point in this story will be grimly familiar to everyone reading this book.

In 2020, disaster struck when a novel coronavirus spread around the world. To anyone who knows the history of pandemics, the devastation caused by COVID-19 was not a surprise. I had been studying disease outbreaks for years as part of my interest in global health, and I had become deeply concerned that the world wasn't ready to handle a pandemic like the 1918 flu, which killed tens of millions of people. In 2015, I had given a TED talk and several interviews in which I made the case that we needed to create a system for detecting and responding to big disease outbreaks. Other people, including former U.S. president George W. Bush, had made similar arguments.

Unfortunately, the world did little to prepare, and when the novel coronavirus struck, it caused massive loss of life and economic pain such as we had not seen since the Great Depression. Although I kept up much of my work on climate change, Melinda and I made COVID-19 the top priority for the Gates Foundation and the main focus of our own work. Every day, I would talk to scientists at universities and small companies, CEOs of pharmaceutical companies, or heads of government to see how the foundation could help accelerate the work on tests, treatments, and vaccines. By November 2020, we had committed more than $445 million in grants to fighting the disease, and hundreds of millions more via various financial investments to get vaccines, tests, and other critical products to lower-income countries faster.

Because economic activity slowed so much, the world emitted fewer greenhouse gases in 2020 than in 2019—as I mentioned earlier, about 4.5 percent less. In real terms, that means we released the equivalent of 50 billion tons of carbon, instead of 52 billion.

That's a meaningful reduction, and we would be in great shape if we could continue that rate of decrease every year. Unfortunately, we can't.

Consider what it took to achieve this 4.5 percent reduction. A million people died, and tens of millions were put out of work. To put it mildly, this was not a situation that anyone would want to continue or repeat. And yet the world's greenhouse gas emissions probably dropped by less than 5 percent, and possibly less than that. What's remarkable to me is not how much emissions went down because of the pandemic, but how little.

This small decline in emissions is proof that we cannot get to zero emissions simply—or even mostly—by flying and driving less. Just as we needed new tests, treatments, and vaccines for the novel coronavirus, we need new tools for fighting climate change: zero-carbon ways to produce electricity, make things, grow food, keep our buildings cool and warm, and move people and goods around the world. And we need new seeds and other innovations to help the world's poorest people—many of whom are smallholder farmers—adapt to a warmer climate.

Of course, there are other hurdles too, and they don't have anything to do with science or funding. In the United States especially, the conversation about climate change has been sidetracked by politics. Some days, it can seem as if we have little hope of getting anything done.

I think more like an engineer than a political scientist, and I don't have a solution to the politics of climate change. Instead, what I hope to do is focus the conversation on what getting to zero requires: We need to channel the world's passion and its scientific IQ into deploying the clean energy solutions we have now, and inventing new ones, so we stop adding greenhouse gases to the atmosphere.

I am aware that I'm an imperfect messenger on climate change. The world is not exactly lacking in rich men with big ideas about what other people should do, or who think technology can fix any problem. And I own big houses and fly in private planes—in fact, I took one to Paris for the climate conference—so who am I to lecture anyone on the environment?

I plead guilty to all three charges.

I can't deny being a rich guy with an opinion. I do believe, though, that it is an informed opinion, and I am always trying to learn more.

I'm also a technophile. Show me a problem, and I'll look for technology to fix it. When it comes to climate change, I know innovation isn't the only thing we need. But we cannot keep the earth livable without it. Techno-fixes are not sufficient, but they are necessary.

Finally, it's true that my carbon footprint is absurdly high. For a long time I have felt guilty about this. I've been aware of how high my emissions are, but working on this book has made me even more conscious of my responsibility to reduce them. Shrinking my carbon footprint is the least that can be expected of someone in my position who's worried about climate change and publicly calling for action.

In 2020, I started buying sustainable jet fuel and will fully offset my family's aviation emissions in 2021. For our non-aviation emissions, I'm buying offsets through a company that runs a facility that removes carbon dioxide from the air (for more on this technology, which is called direct air capture, see chapter 4, "How We Plug In"). I'm also supporting a nonprofit that installs clean energy upgrades in affordable housing units in Chicago. And I'll keep looking for other ways to reduce my personal footprint.

I'm also investing in zero-carbon technologies. I like to think of these as another kind of offset for my emissions. I've put more than $1 billion into approaches that I hope will help the world get to zero, including affordable and reliable clean energy and low-emissions cement, steel, meat, and more. And I'm not aware of anyone who's investing more in direct air capture technologies.

Of course, investing in companies doesn't make my carbon footprint smaller. But if I've picked any winners at all, they'll be responsible for removing much more carbon than I or my family is responsible for. Besides, the goal isn't simply for any one person to make up for his or her emissions; it's to avoid a climate disaster. So I'm supporting early-stage clean energy research, investing in promising clean energy companies, advocating for policies that will trigger breakthroughs throughout the world, and encouraging other people who have the resources to do the same.

Here's the key point: Although heavy emitters like me should use less energy, the world overall should be using *more* of the goods and services that energy provides. There is nothing wrong with using more energy as long as it's carbon-free. The key to addressing climate change is to make clean energy just as cheap and reliable as what we get from fossil fuels. I'm putting a lot of effort into what I think will get us to that point and make a meaningful difference in going from 52 billion tons a year to zero.

—

This book suggests a way forward, a series of steps we can take to give ourselves the best chance to avoid a climate disaster. It breaks down into five parts:

Why zero? In chapter 1, I'll explain more about why we need to get to zero, including what we know (and what we don't) about how rising temperatures will affect people around the world.

The bad news: Getting to zero will be really hard. Because every plan to achieve anything starts with a realistic assessment of the barriers that stand in your way, in chapter 2 we'll take a moment to consider the challenges we're up against.

How to have an informed conversation about climate change. In chapter 3, I'll cut through some of the confusing statistics you might have heard and share the handful of questions I keep in mind in every conversation I have about climate change. They have kept me from going wrong more times than I can count, and I hope they will do the same for you.

The good news: We can do it. In chapters 4 through 9, I'll break down the areas where today's technology can help and where we need breakthroughs. This will be the longest part of the book, because there's so much to cover. We have some solutions we need to deploy in a big way now, and we also need a *lot* of innovations to be developed and spread around the world in the next few decades.

Although I'll introduce you to some of the technologies that I am especially excited about, I'm not going to name many specific companies. Partly that's because I'm investing in some of them, and I don't want to look as if I'm favoring companies that I have a financial interest in. But more important, I want the focus to be on the ideas and innovations, not on particular businesses. Some companies may go under in the coming years; that comes with the territory when you're doing cutting-edge work, though it's not necessarily a sign of failure. The key thing is to learn from the failure and incorporate the lessons into the next venture, just as we did at Microsoft and just as every other innovator I know does.

Steps we can take now. I wrote this book because I see not just the problem of climate change; I also see an opportunity to solve it. That's not pie-in-the-sky optimism. We already have two of the three things you need to accomplish any major undertaking. First, we have ambition, thanks to the passion of a growing global movement led by young people who are deeply concerned about climate change. Second, we have big goals for solving the problem as more national and local leaders around the world commit to doing their part.

Now we need the third component: a concrete plan to achieve our goals.

Just as our ambitions have been driven by an appreciation for climate science, any practical plan for reducing emissions has to be driven by other disciplines: physics, chemistry, biology, engineering, political science, economics, finance, and more. So in the final chapters of this book, I'll propose a plan based on guidance I've gotten from experts in all these disciplines. In chapters 10 and 11, I'll focus on policies that governments can adopt; in chapter 12, I'll suggest steps that each of us can take to help the world get to zero. Whether you're a government leader, an entrepreneur, or a voter with a busy life and too little free time (or all of the above), there are things you can do to help avoid a climate disaster.

That's it. Let's get started.

CHAPTER 1

WHY ZERO?

The reason we need to get to zero is simple. Greenhouse gases trap heat, causing the average surface temperature of the earth to go up. The more gases there are, the more the temperature rises. And once greenhouse gases are in the atmosphere, they stay there for a very long time; something like one-fifth of the carbon dioxide emitted today will still be there in 10,000 years.

There's no scenario in which we keep adding carbon to the atmosphere and the world stops getting hotter, and the hotter it gets, the harder it will be for humans to survive, much less thrive. We don't know exactly how much harm will be caused by a given rise in the temperature, but we have every reason to worry. And, because greenhouse gases remain in the atmosphere for so long, the planet will stay warm for a long time even after we get to zero.

Admittedly, I'm using "zero" imprecisely, and I should be clear about what I mean. In preindustrial times—before the mid-18th century or so—the earth's carbon cycle was probably roughly in balance; that is, plants and other things absorbed about as much carbon dioxide as was emitted.

But then we started burning fossil fuels. These fuels are made of carbon that's stored underground, thanks to plants that died eons ago and got compressed over millions of years into oil, coal, or natural gas. When we dig up those fuels and burn them, we emit extra carbon and add to the total amount in the atmosphere.

There are no realistic paths to zero that involve abandoning these fuels completely or stopping all the other activities that also produce greenhouse gases (like making cement, using fertilizer, or letting methane leak out of natural gas power plants). Instead, in all likelihood, in a zero-carbon future we will still be producing some emissions, but we'll have ways to remove the carbon they emit.

In other words, "getting to zero" doesn't actually mean "zero." It means "near net zero." It's not a pass-fail exam where everything's great if we get a 100 percent reduction and everything's a disaster if we get only a 99 percent reduction. But the bigger the reduction, the bigger the benefit.

A 50 percent drop in emissions wouldn't stop the rise in temperature; it would only slow things down, somewhat postponing but not preventing a climate catastrophe.

And suppose we reach a 99 percent reduction. Which countries and sectors of the economy would get to use the remaining 1 percent? How would we even decide something like that?

In fact, to avoid the worst climate scenarios, at some point we'll not only need to stop adding more gases but actually need to start removing some of the gases we have already emitted. You may see this step referred to as "net-negative emissions." It just means that eventually, we'll need to take more greenhouse gases out of the atmosphere than we put in so that we can limit the temperature increase. To return to the bathtub analogy from the introduction: We won't just shut off the flow of water into the tub. We'll open up the drain and let water flow out too.

I suspect that this chapter will not be the first place you'll have read about the risks of failing to get to zero. After all, climate change is in the news just about every day, as it should be: It's an urgent problem, and it deserves every headline it gets. But the coverage can be confusing and even contradictory.

In this book, I'll try to cut through the noise. Over the years, I've had the chance to learn from some of the world's top climate and energy scientists. It's a never-ending conversation, because researchers' understanding of the climate is always advancing as they incorporate new data and improve the computer models they use to forecast different scenarios. But I've found it enormously helpful in sorting out what's likely from what's possible but not probable, and it has convinced me that the only way to avoid disastrous outcomes is to get to zero. In this chapter I want to share some of what I've learned.

A Little Is a Lot

I was surprised when I learned that what sounded like a small increase in the global temperature—just 1 or 2 degrees Celsius, which is 1.8 to 3.6 degrees Fahrenheit—could actually cause a lot of trouble.* But it's true: In climate terms, a change of just a few degrees is a big deal. During the last ice age, the average temperature was just 6 degrees Celsius lower than it is today. During the age of the dinosaurs, when the average temperature was perhaps 4 degrees Celsius higher than today, there were crocodiles living above the Arctic Circle.

It's also important to remember that these average numbers can

* Most climate change reports use the Celsius scale for reporting temperature changes. I'll follow that practice in this book, because that's what you'll see in most news reports. To get an idea of a temperature change in Fahrenheit that is accurate enough for most purposes, you can just double the Celsius number and remember that your estimate is a little high. Since most Americans think more naturally in Fahrenheit, I'll use that scale when I'm referring to daily temperatures.

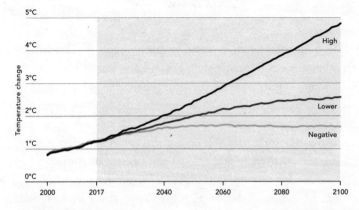

Three lines you should know. These lines show you how much the temperature might go up in the future if emissions grow a lot (that's the high line), if they grow less (lower), and if we start removing more carbon than we emit (negative). (KNMI Climate Explorer)

obscure a pretty big range of temperatures. Even though the global mean has gone up just 1 degree Celsius since preindustrial times, some places have already started experiencing temperature increases of more than 2 degrees Celsius. These regions are home to between 20 percent and 40 percent of the world's population.

Why are some places heating up more than others? In the interior of some continents, the soil is drier, which means the land can't cool off as much as it did in the past. Basically, continents aren't sweating as much as they used to.

So what does a warming planet have to do with greenhouse gas emissions? Let's start with the basics. Carbon dioxide is the most common greenhouse gas, but there are a handful of others, such as nitrous oxide and methane. You might have enjoyed nitrous oxide at the dentist's office—it's also known as laughing gas—and methane is the main ingredient in the natural gas that you might use to run your stove or water heater. Molecule for molecule, many of these other gases cause more warming than carbon dioxide does—in the case of methane, 120 times more warming the moment it reaches the atmosphere. But methane doesn't stay around as long as carbon dioxide.

To keep things simple, most people combine all the different greenhouse gases into a single measure known as "carbon dioxide equivalents." (You might see the term abbreviated as CO_2e.) We use carbon dioxide equivalents to account for the fact that some gases trap more heat than carbon dioxide but don't stay around as long. Unfortunately, carbon dioxide equivalents are an imperfect measure: Ultimately, what really matters isn't the amount of greenhouse gas emissions; what matters is the higher temperatures and their impact on humans. And on that front, a gas like methane is much worse than carbon dioxide. It drives the temperature up immediately, and by quite a bit. When you use carbon dioxide equivalents, you aren't fully accounting for this important short-term effect.

Nevertheless, they're the best method we have for counting emissions, and they come up often in discussions about climate change, so I'll use them in this book. The 52 billion tons I keep mentioning is the world's annual emissions in carbon dioxide equivalents. You may see numbers like 37 billion elsewhere—that's just carbon dioxide, without the other greenhouse gases—or 10 billion, which is just the carbon itself. For the sake of variety, and because reading "greenhouse gas" a hundred times will make your eyes glaze over, I'll sometimes use "carbon" as a synonym for carbon dioxide and the other gases.

Greenhouse gas emissions have increased dramatically since the 1850s due to human activity, such as burning fossil fuels. Check out the charts on page 24. On the left you can see how much our carbon dioxide emissions have grown since 1850, and on the right you can see how much the average global temperature has gone up.

How do greenhouse gases cause warming? The short answer: They absorb heat and trap it in the atmosphere. They work the same way a greenhouse works—hence the name.

You've actually seen the greenhouse effect in action on a very different scale, whenever your car is sitting outside in the sun: Your windshield lets sunlight in, then traps some of that energy. That's why the interior of your car can get so much hotter than the outside temperature.

But that explanation only raises more questions. How can the sun's heat get past greenhouse gases on its way to the earth but then get trapped by these same gases in our atmosphere? Does carbon dioxide work like some giant one-way mirror? For that matter, if carbon dioxide and methane trap heat, why doesn't oxygen?

The answers lie in a neat bit of chemistry and physics. As you may recall from physics class, all molecules vibrate; the faster they vibrate, the hotter they are. When certain types of molecules are hit with radiation at certain wavelengths, they block the radiation, soak up its energy, and vibrate faster.

But not all radiation is on the right wavelengths to cause this effect. Sunlight, for example, passes right through most greenhouse gases without getting absorbed. Most of it reaches the earth and warms up the planet, just as it has been doing for eons.

Here's the problem: The earth doesn't hold on to all that energy forever; if it did, the planet would already be unbearably hot. Instead, it radiates some of the energy back toward space, and some of this energy is emitted in just the right range of wavelengths to get absorbed by greenhouse gases. Rather than going out harmlessly into the void, it hits the greenhouse molecules and makes them vibrate faster, heating up the atmosphere. (By the way, we should be thankful for the greenhouse effect; without it, the planet would be far too cold for us. The problem is that all these extra greenhouse gases are sending the effect into overdrive.)

Why don't all gases act this way? Because molecules with two copies of the same atom—for example, nitrogen or oxygen molecules—let radiation pass straight through them. Only molecules made up of different atoms, the way carbon dioxide and methane are, have the right structure to absorb radiation and start heating up.

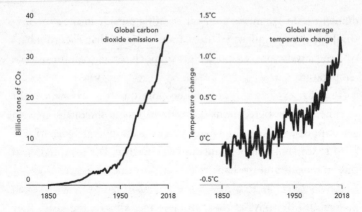

Carbon dioxide emissions are on the rise, and so is the global temperature. On the left you see how our carbon dioxide emissions from industrial processes and burning fossil fuels have gone up since 1850. On the right you see how the global average temperature is rising along with emissions. (Global Carbon Budget 2019; Berkeley Earth)

So that's the first part of the answer to the question "Why do we have to get to zero?"—because every bit of carbon we put into the atmosphere adds to the greenhouse effect. There's no getting around physics.

The next part of the answer involves the impact that all those greenhouse gases are having on the climate, and on us.

What We Do and Don't Know

Scientists still have a lot to learn about how and why the climate is changing. IPCC reports acknowledge up front some uncertainty about how much and how quickly the temperature will go up, for example, and exactly what effect these higher temperatures will have.

One problem is that computer models are far from perfect. The climate is mind-blowingly complex, and there's a lot we don't understand about things like how clouds affect warming or the impact of all this extra heat on ecosystems. Researchers are identifying these gaps and trying to fill them in.

Still, there is a lot that scientists do know and can state with confidence about what will happen if we don't get to zero. Here are a few key points.

The earth is warming, it's warming because of human activity, and the impact is bad and will get much worse. We have every reason to believe that at some point the impact will be catastrophic. Will that point come in 30 years? Fifty years? We don't know precisely. But given how hard the problem will be to solve, even if the worst case is 50 years away, we need to act now.

We've already raised the temperature at least 1 degree Celsius since preindustrial times, and if we don't reduce emissions, we'll probably have between 1.5 and 3 degrees Celsius of warming by mid-century, and between 4 and 8 degrees Celsius by the end of the century.

All this extra heat will cause various changes in the climate. Before I explain what's coming, I have to give you one caveat: Although we can predict the course of broad trends, like "there will be more hot days" and "sea levels will go up," we can't with certainty blame climate change for any particular event. For example, when there's a heat wave, we can't say whether it was caused by climate change alone. What we can do, though, is say how much climate change increased the odds of that heat wave happening. For hurricanes, it's unclear whether warmer oceans are causing a rise in the number of storms, but there is growing evidence that climate change is making storms wetter and increasing the number of intense ones. We also don't know whether or to what extent these extreme events will interact with each other to produce even more serious effects.

What else do we know?

For one thing, there will be more really hot days. I could give you statistics from cities throughout the United States, but I'll pick

Albuquerque, New Mexico, because I have a special connection with the place: It's where Paul Allen and I founded Microsoft in 1975. (Micro-Soft, to be totally accurate—we wisely dropped the hyphen and lowercased the *S* a couple of years later.) In the mid-1970s, when we were just getting started, the temperature in Albuquerque went over 90 degrees Fahrenheit about 36 times a year, on average. By mid-century, the city's thermometers will go over 90 at least twice as often every year. By the end of the century, the city could see as many as 114 days that hot. In other words, they'll go from one month's worth of hot days every year to three months' worth.

Not everyone will suffer equally from hotter and more humid days. For example, the Seattle area, where Paul and I moved Microsoft in 1979, will probably get off relatively easy. We might reach 90 degrees on as many as 14 days a year later this century, after having an average of just one or two a year in the 1970s. And some places might actually benefit from a warmer climate. In cold regions, for example, fewer people will die of hypothermia and the flu, and they'll spend less money to heat their homes and businesses.

But the overall trend points toward trouble from a hotter climate. And all this extra heat has knock-on effects; for instance, it means that storms are getting worse. Scientists are still debating whether storms are happening more often because of the heat, but they appear to be getting more powerful in general. We know that when the average temperature rises, more water evaporates from the earth's surface into the air. Water vapor is a greenhouse gas, but unlike carbon dioxide or methane, it doesn't stay in the air for long—eventually, it falls back to the surface as rain or snow. As water vapor condenses into rain, it releases a massive amount of energy, as anyone who has ever experienced a big thunderstorm knows.

Even the most powerful storm typically lasts only a few days, but its impact can reverberate for years. There's the loss of life, a tragedy in its own right that can leave survivors both heartbroken and, often, destitute. Hurricanes and floods also destroy buildings, roads, and power lines that took years to build. All of that property can eventually be replaced, of course, but doing so siphons off money and time that could be put into new investments that help the economy grow. You're always trying to catch up to where you were, instead of getting ahead. One study estimated that Hurricane Maria in 2017 set Puerto Rico's infrastructure back more than two decades. How long before the next storm comes along and sets it back again? We don't know.

Hurricane Maria set Puerto Rico's power grid and other infrastructure back some two decades, according to one study.

These stronger storms are creating a strange feast-or-famine situation: Even though it's raining more in some places, other places are experiencing more frequent and more severe droughts. Hotter air can hold more moisture, and as the air gets warmer, it gets thirstier, drinking up more water from the soil. By the end of the century, soils in the southwestern United States will have 10 percent to 20 percent less moisture, and the risk of drought there will go up by at least 20 percent. Droughts will also threaten the Colorado River, which supplies drinking water for nearly 40 million people and irrigation for more than one-seventh of all American crops.

A hotter climate means there will be more frequent and destructive wildfires. Warm air absorbs moisture from plants and soil, leaving everything more prone to burning. There's a lot of variation around the world, because conditions change so much from place to place. But California is a dramatic example of what's going on. Wildfires now occur there five times more often than in the 1970s, largely because the fire season is getting longer and the forests there now contain much more dry wood that's likely to burn. According to the U.S. government, half of this increase is due to climate change, and by mid-century America could experience more than twice as much destruction from wildfires as it does today. This should be worrisome for anyone who remembers America's devastating wildfire season of 2020.

Another effect of the extra heat is that sea levels will go up. This is partly because polar ice is melting, and partly because seawater expands when it gets warmer. (Metal does the same thing, which is why you can loosen a ring that's stuck on your finger by running it under hot water.) Although the overall rise in the global average sea level—most likely, a few feet by 2100—may not sound like much, the rising tide will affect some places much more than others. Beach areas are in trouble, not surprisingly, but so are cities situated on especially porous land. Miami is already seeing seawater bubble up through storm drains, even when it isn't raining—that's called dry-weather flooding—and the situation will not get better. In the IPCC's moderate scenario, by 2100 the sea level around Miami will rise by nearly two feet. And some parts of the city are settling—sinking, essentially—which might add another foot of water on top of that.

Rising sea levels will be even worse for the poorest people in the world. Bangladesh, a poor country that's making good progress on the path out of poverty, is a prime example. It has always been beset by severe weather; it has hundreds of miles of coastline on the Bay of Bengal; most of the country sits in low-lying, flood-prone river deltas; and it gets heavy rainfall every year. But the changing climate is making life there even harder. Thanks to cyclones, storm surges, and river floods, it is now common for 20 to 30 percent of Bangladesh to be underwater, wiping out crops and homes and killing people throughout the country.

Finally, with the extra heat and the carbon dioxide that's causing it, plants and animals are being affected. According to research cited by the IPCC, a rise of 2 degrees Celsius would cut the geographic range of vertebrates by 8 percent, plants by 16 percent, and insects by 18 percent.

For the food we eat, it's a mixed picture, though mainly a grim one. On the one hand, wheat and many other plants grow faster and need less water when there's a large amount of carbon in the air. On the other hand, corn is especially sensitive to heat, and it's the number one crop in the United States, worth more than $50 billion a year. In Iowa alone, more than 13 million acres of land are planted with corn.

Globally, there's a wide range of possibilities for how climate change could affect the amount of food we get from each acre of crops. In some northerly regions, yields could go up, but in most places they'll drop, by anywhere from a few percentage points to as much as 50 percent. Climate change could cut southern Europe's production of wheat and corn in half by mid-century. In sub-Saharan Africa, farmers could see the growing season shrink by 20 percent and millions of acres of land become substantially drier. In poor communities, where many people already spend more than half of their incomes on food, food prices could rise by 20 percent or more. Extreme droughts in China—whose agricultural system provides wheat, rice, and corn for a fifth of the world's population—could trigger a regional or even global food crisis.

Extra heat won't be good for the animals we eat and get milk from; it will make them less productive and more prone to dying young, which in turn will make meat, eggs, and dairy more expensive. Communities that rely on seafood will have trouble too, because not only are the seas getting warmer, they're also bifurcating—developing some places where the water has more oxygen and others where it has less oxygen. As a result, fish and other sea life are moving to different waters, or simply dying off. If the temperature rises by 2 degrees Celsius, coral reefs could vanish completely, destroying a major source of seafood for more than a billion people.

When It Doesn't Rain, It Pours

You might think that the difference between 1.5 and 2 degrees would not be that great, but climate scientists have run simulations of both scenarios, and the news is not good. In many ways, a 2-degree rise wouldn't simply be 33 percent worse than 1.5; it could be 100 percent worse. Twice as many people would have trouble getting clean water. Corn production in the tropics would go down twice as much.

Any one of these effects of climate change will be bad enough. But no one's going to suffer from just hot days, or just floods, and nothing else. That's not how climate works. The effects of climate change add up, one on top of the other.

As it gets hotter, for example, mosquitoes will start living in new places (they like it humid, and they'll move from areas that dry out to ones that become more humid), so we'll see cases of malaria and other insect-borne diseases where they've never appeared before.

Heatstroke will be another major problem, and it's linked to the humidity, of all things. Air can contain only a certain amount of water vapor, and at some point it hits a ceiling, becoming so saturated that it can't absorb any more moisture. Why does that matter? Because the human body's ability to cool off depends on the air's ability to absorb sweat as it evaporates. If the air can't absorb your sweat, then it can't cool you off, no matter how much you perspire. There's simply nowhere for your perspiration to go. Your body temperature stays high, and if nothing changes, you die of heatstroke within hours.

Heatstroke, of course, is nothing new. But as the atmosphere gets hotter and more humid, it will become a much bigger problem. In the regions that are most in jeopardy—the Persian Gulf, South Asia, and parts of China—there will be times of the year when hundreds of millions of people will be at risk of dying.

To see what happens when these effects start piling up, let's look at the impact on individual people. Imagine you're a prosperous young farmer raising corn, soybeans, and cattle in Nebraska in 2050. How might climate change affect you and your family?

You're in the middle of the United States, far from the coasts, so rising sea levels don't directly harm you. But the heat does. In the 2010s, when you were a kid, you might've seen 33 days a year when the temperature hit 90; now it happens 65 or 70 times a year. The rain is also a lot less reliable: When you were a kid, you could expect around 25 inches a year; now it might be as little as 22 or as much as 29.

Maybe you've adjusted your business to the hotter days and the unpredictable rain. Years ago, you invested in new crop varieties that can tolerate extra heat, and you've found work-arounds that let you stay inside during the worst part of the day. You didn't love spending extra money on these crops and work-arounds, but they're better than the alternative.

One day, a powerful storm strikes without warning. As nearby rivers spill over the levees that have held them back for decades, your farm gets flooded. It's the kind of deluge your parents would've called a hundred-year flood, but now you'd consider yourself lucky if it happened only once a decade. The waters wash away large portions of your corn and soybean crops, and your stored grain is soaked so thoroughly that it rots and you have to throw it away. In theory, you could sell your cattle to make up for the loss, but all your cattle feed has been swept away too, so you won't be able to keep them alive for long.

Eventually the waters recede, and you can see that the nearby roads, bridges, and rail lines are now unusable. Not only does that keep you from shipping out whatever grain you've managed to preserve; it also makes it harder for trucks to deliver the seeds you need for the next planting season, assuming your fields are still usable. It all adds up to a disaster that could end your farming career and force you to sell off land that has been in your family for generations.

It may sound as if I'm cherry-picking the most extreme example, but things like this are already happening, especially to poor farmers, and in a few decades they could be happening to far more people. And as bad as it sounds, if you take a global perspective, you'll see that things will be much worse for the poorest 1 billion people in the world—people who are already struggling to get by and who will only struggle more as the climate gets worse.

Now imagine you live in rural India, where you and your husband are subsistence farmers, which means you and your kids eat nearly all the food you raise. In an especially good season, you sometimes have enough left over to sell so you can buy medicine for your kids or send them to school. Unfortunately, heat waves have become so common that your village is becoming unlivable—it's not at all unusual to have several days in a row over 120 degrees—and between the heat and the pests that are now invading your fields for the first time, it's almost impossible to keep your crops alive. Although monsoons have flooded other parts of the country, your community has received far less rain than normal, making it so hard to find water that you survive off a trickling pipe that runs only a few times a week. It's getting even tougher to simply keep your family fed.

You've already sent your oldest son to work in a big city hundreds of miles away because you couldn't afford to feed him. One of your neighbors committed suicide when he couldn't support his family anymore. Should you and your husband stay and try to survive on the farm you know, or abandon the land and move to a more urban area where you might make a living?

It's a wrenching decision. But it's the kind of choice that people around the world are already facing, with heartbreaking results. In the worst drought ever recorded in Syria—which lasted from 2007 to 2010—some 1.5 million people left farming areas for the cities, helping to set the stage for the armed conflict that started in 2011. That drought was made three times more likely by climate change. By 2018, roughly 13 million Syrians had been displaced.

This problem is only going to get worse. One study that looked at the relationship between weather shocks and asylum applications to the European Union found that even with moderate warming, asylum applications could go up by 28 percent, to nearly 450,000 a year, by the end of the century. The same study estimated that by 2080 lower crop yields would cause between 2 percent and 10 percent of adults in Mexico to try to cross the border into the United States.

Let's put all this into terms that everyone who is experiencing the COVID-19 pandemic can relate to. If you want to understand the kind of damage that climate change will inflict, look at COVID-19 and then imagine spreading the pain out over a much longer period of time. The loss of life and economic misery caused by this pandemic are on par with what will happen regularly if we do not eliminate the world's carbon emissions.

I'll start with the loss of life. How many people will be killed by COVID-19 versus by climate change? Because we want to compare events that happen at different points in time—the pandemic in

2020 and climate change in, say, 2030—and the global population will change in that time, we can't compare the absolute numbers of deaths. Instead we will use the death rate: that is, the number of deaths per 100,000 people.

Using data from the Spanish flu of 1918 and the COVID-19 pandemic and averaging it out over the course of a century, we can estimate the amount by which a global pandemic increases the global mortality rate. It's about 14 deaths per 100,000 people each year.

How does that compare to climate change? By mid-century, increases in global temperatures are projected to raise global mortality rates by the same amount—14 deaths per 100,000. By the end of the century, if emissions growth stays high, climate change could be responsible for 75 extra deaths per 100,000 people.

In other words, by mid-century, climate change could be just as deadly as COVID-19, and by 2100 it could be five times as deadly.

The economic picture is also bleak. The likely impacts from climate change and from COVID-19 vary quite a bit, depending on which economic model you use. But the conclusion is unmistakable: In the next decade or two, the economic damage caused by climate change will likely be as bad as having a COVID-sized pandemic every 10 years. And by the end of the 21st century, it will be much worse if the world remains on its current emissions path.*

Many of the predictions in this chapter may sound familiar to you if you've been following climate change in the news. But as the

---

* Here's the math. Recent models suggest that the cost of climate change in 2030 will likely be between 0.85 percent and 1.5 percent of America's GDP per year. Meanwhile, current estimates for the cost of COVID-19 to the United States this year range between 7 percent and 10 percent of GDP. If we assume that a similar disruption happens once every 10 years, that's an average annual cost of 0.7 percent to 1 percent of GDP—roughly equivalent to the projected damage from climate change.

temperature goes up, all these problems will happen more often, more severely, and to more people. And there's a chance of relatively sudden catastrophic climate change, if, for example, large sections of the earth's permanently frozen ground (called permafrost) gets warm enough to melt and releases the huge amounts of greenhouse gases, mostly methane, that are trapped there.

Despite the scientific uncertainties that remain, we understand enough to know that what's coming will be bad. There are two things we can do about it:

Adaptation. We can try to minimize the impact of the changes that are already here and that we know are coming. Because climate change will have the worst impact on the world's poorest people, and most of the world's poorest people are farmers, adaptation is a major focus for the agriculture team at the Gates Foundation. For example, we're funding a lot of research into new varieties of crops that tolerate the droughts and floods that will be more frequent and severe in the coming decades. I'll explain more about adaptation and outline a few of the steps we'll need to take in chapter 9.

Mitigation. Most of this book isn't about adaptation. It's about the other thing we need to do: stop adding greenhouse gases to the atmosphere. To have any hope of staving off disaster, the world's biggest emitters—the richest countries—have to get to net-zero emissions by 2050. Middle-income countries need to get there soon after, and the rest of the world will eventually need to follow suit.

I've heard people object to the idea that rich countries should go first: "Why should we bear the brunt of this?" It's not simply because we've caused most of the problem (although that's true). It's also because this is a huge economic opportunity: The countries that build great zero-carbon companies and industries will be the ones that lead the global economy in the coming decades.

Rich countries are best suited to develop innovative climate solutions; they're the ones with government funding, research universities, national labs, and start-up companies that draw talent from all over the world, so they'll need to lead the way. Whoever makes big energy breakthroughs and shows they can work on a global scale, and be affordable, will find many willing customers in emerging economies.

I see many different pathways that can get us to zero. Before exploring them in detail, we need to take stock of just how hard the journey will be.

CHAPTER 2

THIS WILL BE HARD

Please don't let the title of this chapter depress you. I hope it's clear by now that I believe we can get to zero, and in the coming chapters I will try to give you a sense of why I feel that way and what it will take to get there. But we can't solve a problem like climate change without an honest accounting of how much we need to do and what obstacles we need to overcome. So with the idea in mind that we will get to solutions—including ways to speed up the transition from fossil fuels—let's look at the biggest barriers we're facing.

Fossil fuels are like water. I'm a big fan of the late writer David Foster Wallace. (I'm preparing for his mammoth novel *Infinite Jest* by slowly making my way through everything else he ever wrote.) When Wallace gave a now-famous commencement speech at Kenyon College in 2005, he started with this story:

> There are these two young fish swimming along, and they happen to meet an older fish swimming the other way, who nods at them and says, "Morning, boys, how's the water?" And the two young fish swim on for a bit, and then eventually

one of them looks over at the other and goes, "What the hell is water?"*

Wallace explained, "The immediate point of the fish story is that the most obvious, ubiquitous, important realities are often the ones that are the hardest to see and talk about."

Fossil fuels are like that. They're so pervasive that it can be hard to grasp all the ways in which they—and other sources of greenhouse gases—touch our lives. I find it helpful to start with everyday objects and go from there.

Did you brush your teeth this morning? The toothbrush probably contains plastic, which is made from petroleum, a fossil fuel.

If you ate breakfast, the grains in your toast and cereal were grown with fertilizer, which releases greenhouse gases when it's made. They were harvested by a tractor that was made of steel—which is made with fossil fuels in a process that releases carbon—and ran on gasoline. If you had a burger for lunch, as I do occasionally, raising the beef caused greenhouse gas emissions—cows burp and fart methane—and so did growing and harvesting the wheat that went into the bun.

If you got dressed, your clothes might contain cotton—also fer-

* You can find the whole speech, "This Is Water," online and in book form. It's wonderful.

tilized and harvested—or polyester, made from ethylene, which is derived from petroleum. If you've used toilet paper, that's more trees cut down and carbon emitted.

If the vehicle you took to work or school today was powered by electricity, great—though that electricity was probably generated using a fossil fuel. If you took a train, it went along tracks made of steel and through tunnels made using cement, which is produced with fossil fuels in a process that releases carbon as a by-product. The car or bus you took is made of steel and plastic. The same goes for the bike you rode last weekend. The roads you drove on contain cement as well as asphalt, which is derived from petroleum.

If you live in an apartment building, you're probably surrounded by cement. If you live in a house made of wood, the lumber was cut and trimmed by gas-powered machines that were made with steel and plastic. If your home or office has heating or air-conditioning, not only is it using a fair amount of energy, but the coolant in the air conditioner may be a potent greenhouse gas. If you're sitting in a chair made of metal or plastic, that's more emissions.

Also, virtually all of these items, from the toothbrush to the building materials, were transported from someplace else on trucks, airplanes, trains, and ships, all of which were themselves powered by fossil fuels and made using fossil fuels.

In other words, fossil fuels are everywhere. Take oil as just one example: The world uses more than 4 billion gallons every day. When you're using any product at that kind of volume, you can't simply stop overnight.

What's more, there's a very good reason why fossil fuels are everywhere: They're so inexpensive. As in, *oil is cheaper than a soft drink*. I could hardly believe this the first time I heard it, but it's true. Here's the math: A barrel of oil contains 42 gallons; the average price in 2021 was around $70 per barrel, so that comes to about $1.67 per gallon. Meanwhile, Costco sells 8 liters of soda for $6, a price that amounts to $2.85 a gallon.

Even after you account for fluctuations in the price of oil, the conclusion is the same: Every day, people around the world rely on more than 4 billion gallons of a product that costs less than Diet Coke.

It's no accident that fossil fuels are so cheap. They're abundant and easy to move. We've created big global industries devoted to drilling for them, processing and moving them, and developing innovations that keep their prices low. And their prices don't reflect the damage they cause—the ways they contribute to climate change, pollution, and environmental degradation when they're extracted and burned. We'll explore this problem in more detail in chapter 10.

Just thinking about the scope of this problem can be dizzying. But it does not need to be paralyzing. By deploying the clean and renewable sources we already have while also making breakthroughs in zero-carbon energy, we can figure out how to reduce our net emissions to zero. The key will be to make the clean approach as cheap—or almost as cheap—as the current technology.

We need to hurry up, though, because . . .

It's not just the rich world. Almost everywhere, people are living longer and healthier lives. Standards of living are going up. There is rising demand for cars, roads, buildings, refrigerators, computers, and air conditioners and the energy to power them all. As a result, the amount of energy used per person will go up, and so will the amount of greenhouse gases emitted per person. Even building the infrastructure we'll need to create all this energy—the wind turbines, solar panels, nuclear plants, electricity storage facilities, and so on— will itself involve releasing more greenhouse gases.

But it's not just that each person will be using more energy; there will also be more people. The global population is headed toward 10 billion by the end of the century, and much of this growth is happening in cities that are highly carbon intensive. The speed of urban growth is mind-boggling: By 2060, the world's building stock—a

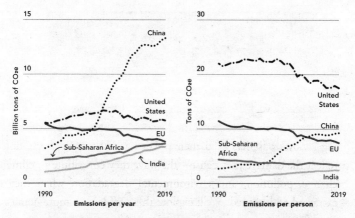

Where the emissions are. Emissions from advanced economies like the United States and Europe have stayed pretty flat or even dropped, but many developing countries are growing fast. That's partly because richer countries have outsourced emissions-heavy manufacturing to poorer ones. (UN Population Division; Rhodium Group)

measure that factors in the number of buildings and their size—will double. That's like putting up another New York City every month for 40 years, and it's mainly because of growth in developing countries like China, India, and Nigeria.

This is good news for every person whose life improves, but it's bad news for the climate we all live in. Consider that nearly 40 percent of the world's emissions are produced by the richest 16 percent of the population. (And that's not counting the emissions from products that are made someplace else but consumed in rich countries.) What will happen as more people live like the richest 16 percent? Global energy demand will go up 50 percent by 2050, and if nothing else changes, carbon emissions will go up by nearly as much. Even if the rich world could magically get to zero today, the rest of the world would still be emitting more and more.

It would be immoral and impractical to try to stop people who are lower down on the economic ladder from climbing up. We can't expect poor people to stay poor because rich countries emitted too many greenhouse gases, and even if we wanted to, there would be no way to accomplish it. Instead, we need to make it possible for

The world will be building the equivalent of another
New York City every month for the next 40 years.

low-income people to climb the ladder without making climate
change worse. We need to get to zero—producing even more
energy than we do today, but without adding any carbon to the
atmosphere—as soon as possible.

Unfortunately . . .

History is not on our side. Judging only by how long previous
transitions have taken, "as soon as possible" is a long time away.

Many farmers still have to use ancient techniques, which is one of the reasons they're trapped in poverty. They deserve modern equipment and approaches, but right now using those tools means producing more greenhouse gases.

We have done things like this before—moving from relying on one energy source to another—and it has always taken decades upon decades. (The best books I have read on this topic are Vaclav Smil's *Energy Transitions* and *Energy Myths and Realities,* which I'm borrowing from here.)

For most of human history, our main sources of energy were our own muscles, animals that could do things like pull plows, and plants that we burned. Fossil fuels did not represent even half of the world's energy consumption until the late 1890s. In China, they didn't take over until the 1960s. There are parts of Asia and sub-Saharan Africa where this transition still hasn't happened.

And consider how long it took for oil to become a big part of our energy supply. We started producing it commercially in the 1860s.

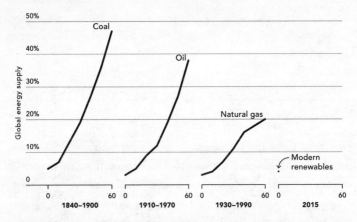

It takes a really long time to adopt new sources of energy. Notice how in 60 years coal went from 5 percent of the world's energy supply to nearly 50 percent. But natural gas reached only 20 percent in the same amount of time. (Vaclav Smil, *Energy Transitions*)

Half a century later, it represented just 10 percent of the world's energy supply. It took 30 years more to reach 25 percent.

Natural gas followed a similar trajectory. In 1900, it accounted for 1 percent of the world's energy. It took seventy years to reach 20 percent. Nuclear fission went faster, going from 0 to 10 percent in 27 years.

This chart shows how much various energy sources grew over the course of 60 years, starting from the time they were introduced. Between 1840 and 1900, coal went from 5 percent of the world's energy supply to nearly 50 percent. But in the 60 years from 1930 to 1990, natural gas reached just 20 percent. In short, energy transitions take a long time.

Fuel sources aren't the only issue. It also takes us a long time to adopt new types of vehicles. The internal combustion engine was introduced in the 1880s. How long before half of all urban families had a car? Thirty to 40 years in the United States, and 70 to 80 years in Europe.

What's more, the energy transition we need now is being driven by something that has never mattered before. In the past, we've moved from one source to another because the new one was cheaper and more powerful. When we stopped burning so much wood and started using more coal, for example, it was because we could get a lot more heat and light from a pound of coal than from a pound of wood.

Or take a more recent example in the United States: We're using more natural gas and less coal to generate electricity. Why? Because new drilling techniques made it much cheaper. It was a matter of economics, not the environment. In fact, whether natural gas is better or worse than coal depends on the way carbon dioxide equivalents are calculated. Some scientists have argued that gas can actually be worse for climate change than coal is, depending on how much leaks out while it's being processed.

Over time, we would naturally start using more renewables, but left to its own devices, this growth won't happen nearly fast enough, and as we'll see in chapter 4, without innovation it won't be enough to get us all the way to zero. We have to force an unnaturally speedy transition. That introduces a level of complexity—in public policy and technology—that we've never had to deal with before.

Why do energy transitions take so long, anyway? Because . . .

Coal plants are not like computer chips. You have probably heard of Moore's Law, the prediction made by Gordon Moore in 1965 that microprocessors would double in power every two years. Gordon turned out to be right, of course, and Moore's Law is one of the main reasons the computing and software industries took off the way they did. As processors got more powerful, we could write better software, which drove up demand for computers, which gave hardware companies the incentive to keep improving their machines, for which we kept writing better software, and on and on in a positive feedback loop.

Moore's Law works because the companies keep finding new ways to make transistors—the tiny switches that power a computer—smaller and smaller. This allows them to pack more transistors onto each chip. A computer chip made today has roughly one million times more transistors on it than one made in 1970, making it a million times more powerful.

You'll sometimes hear Moore's Law invoked as a reason to think we can make the same kind of exponential progress on energy. If computer chips can improve so much so quickly, can't cars and solar panels?

Unfortunately, no. Computer chips are an outlier. They get better because we figure out how to cram more transistors on each one, but there's no equivalent breakthrough to make cars use a million times less gas. Consider that the first Model T that rolled off Henry Ford's production line in 1908 got no better than 21 miles to the gallon. As I write this, the top hybrid on the market gets 58 miles to the gallon. In more than a century, fuel economy has improved by less than a factor of three.

Nor have solar panels become a million times better. When crystalline silicon solar cells were introduced in the 1970s, they converted about 15 percent of the sunlight that hit them into electricity. Today they convert around 25 percent. That's good progress, but it's hardly in line with Moore's Law.

Technology is only one reason that the energy industry can't change as quickly as the computer industry. There's also size. The energy industry is simply enormous—at around $5 trillion a year, one of the biggest businesses on the planet. Anything that big and complex will resist change. And consciously or not, we have built a lot of inertia into the energy industry.

For context, think about how the software business operates. There's no regulatory agency that has to approve your products. Even if you release a piece of software that's imperfect, your customers can still get enthusiastic and give you feedback about how to make it better, as long as the net benefit you're offering is high enough. And virtually all your costs are up front. After you've developed a product, the marginal cost of making more of it is close to zero.

Compare that with the drug and vaccine industry. Getting a new medicine to market is much harder than releasing a new piece of software. Which is as it should be, considering that a drug that makes people sick is much worse than an app that has some flaws. Between basic research, drug development, regulatory approval to test the drug, and every other step required, it takes years for a new medicine to reach patients. But once you have a pill that works, it's very cheap to make more of it.

Now compare both with the energy industry. First, you have huge capital costs that never go away. If you spend $1 billion building a coal plant, the next plant you build will not be any cheaper. And your investors put up that money with the expectation that the plant will run for 30 years or more. If someone comes along with a better technology 10 years down the road, you're not going to just shut down your old plant and go build a new one. At least not without a very good reason—like a big financial payoff, or government regulations that force you to.

Society also tolerates very little risk in the energy business, understandably so. We demand reliable electricity; the lights had better come on every time a customer flips a switch. We also worry about disasters. In fact, safety concerns have nearly killed off new construction of nuclear plants in the United States. Since the accidents at Three Mile Island and Chernobyl, America has broken ground on just two nuclear plants, even though more people die from coal pollution in a single year than have died in all nuclear accidents combined.

We have a large and understandable incentive to stick with what we know, even if what we know is killing us. What we need to do is change the incentives so that we can build an energy system that is all the things we like (reliable, safe) and none of the things we don't like (dependent on fossil fuels). But that will not be easy, because . . .

Our laws and regulations are so outdated. The phrase "government policy" doesn't exactly set people's hair on fire. But policies—everything from tax rules to environmental regulations—have a huge impact on how people and companies behave. We won't get to zero unless we get this right, and we're a long way from doing that. (I'm talking here about the United States, but this applies to many other countries too.)

One problem is that many of the environmental laws and regulations in place today weren't designed with climate change in mind. They were adopted to solve other problems, and now we're trying to use them to reduce emissions. We might as well try to create artificial intelligence using a 1960s mainframe computer.

For example, America's best-known law related to air quality, the Clean Air Act, barely mentions greenhouse gases at all. That's hardly surprising, because it was originally passed in 1970 to reduce the health risks from local air pollution, not to deal with rising temperatures.

Or consider the fuel-economy standards known as CAFE (Corporate Average Fuel Economy). They were adopted in the 1970s because oil prices were skyrocketing and Americans wanted more fuel-efficient cars. Fuel efficiency is great, but now we need to put more electric vehicles on the road, and CAFE standards haven't helped much at all with that, because they weren't designed to.

Outdated policies are not the only problem. Our approach to climate and energy keeps changing with the election cycle. Every four to eight years, a new administration arrives in Washington with its own energy priorities. There's nothing inherently wrong with changing priorities—it happens throughout the government with every new administration—but it takes a toll on researchers who depend on the government for grant money and entrepreneurs who rely on tax incentives. It's hard to make real progress if every few years you have to stop work on one project and start from scratch on something else.

The election cycle also creates uncertainty in the private market. The government offers various tax breaks designed to get more companies to work on clean energy breakthroughs. But they're of limited use, because energy innovation is so hard and can take decades to come to fruition. You could work on an idea for years, only to see a new administration come in and eliminate the incentive you've been counting on.

The bottom line is that our current energy policies will have only a negligible impact on future emissions. You can measure their effect by adding up the extent to which emissions will go down by the year 2030 as a result of all the federal and state policies now on the books. All told, it comes to about 300 million tons, or about 5 percent of projected U.S. emissions in 2030. That's nothing to scoff at, but it's not going to be enough to get us near zero.

Which is not to say that we can't come up with policies that make a big difference on emissions. CAFE standards and the Clean Air Act did what they were designed to do: Cars got more efficient, and the air got cleaner. And there are some effective emissions-related policies in place now, although they're disconnected from each other and don't add up to enough to make a real difference for the climate problem.

I believe that we can do this, but it will be hard. For one thing, it's much easier to tinker with an existing law than to introduce a major new one. It takes a long time to develop a new policy, get public input, go through the court system if there's a legal challenge, and finally implement it. Not to mention the fact that . . .

There isn't as much of a climate consensus as you might think. I'm not talking about the 97 percent of scientists who agree that the climate is changing because of human activities. It's true that there are still small but vocal—and, in some cases, politically powerful— groups of people who are not persuaded by the science. But even if you accept the fact of climate change, you don't necessarily buy the idea that we should be investing large amounts of money in breakthroughs designed to deal with it.

For example, some people argue, *Yes, climate change is happening, but it's not worth spending much to try to stop it or adapt to it. Instead, we should prioritize other things that have a bigger impact on human welfare, like health and education.*

Here's my reply to that argument: Unless we move fast toward zero, bad things (and probably many of them) will happen well within most people's lifetime, and very bad things will happen within a generation. Even if climate change doesn't rank as an existential threat to humanity, it will make most people worse off, and it will make the poorest even poorer. It will keep getting worse until we stop adding greenhouse gases to the atmosphere, and it deserves to be as much of a priority as health and education.

Another argument you often hear goes like this: *Yes, climate change is real, and its effects will be bad, and we have everything we need to stop it. Between solar power, wind power, hydropower, and a few other tools, we're good. It's simply a matter of having the will to deploy them.*

Chapters 4 through 8 explain why I don't buy that notion. We have some of what we need, but far from all of it.

There's another challenge to building a climate consensus: Global cooperation is notoriously difficult. It's hard to get every country in the world to agree on anything—especially when you're asking them to incur some new cost, like the expense of curbing carbon emissions. No single country wants to pay to mitigate its emissions unless everyone else will too. That's why the Paris Agreement, in which more than 190 countries signed up to eventually limit their emissions, was such an achievement. Not because the current commitments will make a huge dent in emissions—if everyone meets them, they'll reduce annual emissions by 3 billion to 6 billion tons in 2030, less than 12 percent of total emissions today—but because it was a starting point that proved global cooperation is possible. The U.S. withdrawal from the 2015 Paris Agreement—a step that President-elect Joe Biden promised to reverse—only illustrates that it's as hard to maintain global compacts as it is to create them in the first place.

To sum up: We need to accomplish something gigantic we have never done before, much faster than we have ever done anything similar. To do it, we need lots of breakthroughs in science and engineering. We need to build a consensus that doesn't exist and create public policies to push a transition that would not happen otherwise. We need the energy system to stop doing all the things we don't like and keep doing all the things we do like—in other words, to change completely and also stay the same.

But don't despair. We can do this. There are lots of ideas out there for how to do it, some of them more promising than others. In the next chapter, I'll explain how I try to tell them apart.

CHAPTER 3

FIVE QUESTIONS TO ASK IN EVERY CLIMATE CONVERSATION

When I started learning about climate change, I kept encountering facts that were hard to get my head around. For one thing, the numbers were so large that they were hard to picture. Who knows what 52 billion tons of gas looks like?

Another problem was that the data I was seeing often appeared devoid of any context. One article said that an emissions-trading program in Europe had reduced the carbon footprint of the aviation sector there by 17 million tons per year. Seventeen million tons certainly sounds like a lot, but is it? What percentage of the total does it represent? The article didn't say, and that kind of omission was surprisingly common.

Eventually, I built a mental framework for the things I was learning. It gave me a sense of how much was a lot and how much was a little, and how expensive something might be. It helped me sort out the most promising ideas. I've found that this approach helps with almost any new topic I'm digging into: I try to get the big picture first, because that gives me the context to understand new information. I'm also more likely to remember it.

The framework of five questions that I came up with still comes in handy today, whether I'm hearing an investment pitch from an energy company or talking with a friend over barbecue in the backyard. Sometime soon you may read an editorial proposing some climate fix; you'll certainly hear politicians touting their plans for climate change. These are complex subjects that can be confusing. This framework will help you cut through the clutter.

1. How Much of the 52 Billion Tons Are We Talking About?

Whenever I read something that mentions some amount of greenhouse gases, I do some quick math, converting it into a percentage of the annual total of 52 billion tons. To me, this makes more sense than the other comparisons you often see, like "this many tons is equivalent to taking one car off the road." Who knows how many cars are on the road to begin with? Or how many cars we would have to take off the road to deal with climate change?

I prefer to connect everything back to the main goal of eliminating 52 billion tons a year. Consider the aviation example I mentioned at the start of this chapter, the program that's getting rid of 17 million tons a year. Divide it by 52 billion and turn it into a percentage. That's a reduction of about 0.03 percent of annual global emissions.

Is that a meaningful contribution? That depends on the answer to this question: Is the number likely to go up, or is it going to stay the same? If this program is starting at 17 million tons but has the potential to reduce emissions by much more, that's one thing. If it's going to stay forever at 17 million tons, that's another. Unfortunately, the answer isn't always obvious. (It wasn't obvious to me when I read about the aviation program.) But it's an important question to ask.

At Breakthrough Energy, we fund only technologies that could remove at least 500 million tons a year if they're successful and fully implemented. That's roughly 1 percent of global emissions. Technologies that will never exceed 1 percent shouldn't compete for the limited resources we have for getting to zero. There may be other good reasons to pursue them, but significantly reducing emissions won't be one of them.

Incidentally, you might have seen references to gigatons of greenhouse gases. A gigaton is a billion tons (or 10^9 tons if you prefer scientific notation). I don't think most people intuitively get what a gigaton of gas is, and besides, eliminating 52 gigatons sounds easier than 52 billion tons, even though they're the same thing. I'll stick with billions of tons.

Tip: Whenever you see some number of tons of greenhouse gases, convert it to a percentage of 52 billion, which is the world's current yearly total emissions (in carbon dioxide equivalents).

2. What's Your Plan for Cement?

If you're talking about a comprehensive plan for tackling climate change, you need to consider everything that humans do to cause greenhouse gas emissions. Some things, like electricity and cars, get lots of attention, but they're only the beginning. Passenger cars represent less than half of all the emissions from transportation, which in turn is 16 percent of all emissions worldwide.

Meanwhile, making steel and cement alone accounts for around 10 percent of all emissions. So the question "What's your plan for cement?" is just a shorthand reminder that if you're trying to come up with a comprehensive plan for climate change, you have to account for much more than electricity and cars.

Here's a breakdown of all the human activities that produce greenhouse gases. Not everyone uses these exact categories, but this is the breakdown I've found most helpful, and it's also the one that the team at Breakthrough Energy uses.*

Getting to zero means zeroing out every one of these categories:

How much greenhouse gas is emitted by the things we do?

Making things (cement, steel, plastic)	29%
Plugging in (electricity)	26%
Growing things (plants, animals)	22%
Getting around (planes, trucks, cargo ships)	16%
Keeping warm and cool (heating, cooling, refrigeration)	7%

You might be surprised to see that making electricity accounts for just over a quarter of all emissions. I know I was taken aback when I learned this: Because most of the articles I read about climate change focused on electricity generation, I assumed it must be the main culprit.

The good news is that even though electricity is only 26 percent of the problem, it could represent much more than 26 percent of the solution. With clean electricity, we could shift away from burning hydrocarbons (which emits carbon dioxide) for fuel. Think electric cars and buses; electric heating and cooling systems in our homes and businesses; and energy-intensive factories using electricity instead of natural gas to make their products. On its own, clean electricity won't get us to zero, but it will be a key step.

---

* These percentages represent global greenhouse gas emissions. When you're categorizing emissions from various sources, one of the questions you have to decide is how to count products that cause emissions both when you make them and when you use them. For example, we produce greenhouse gases when we refine oil into gasoline and again when we burn the gasoline. In this book, I've included all the emissions from making things in "How we make things" and all the emissions from using them in their respective categories. So oil refining goes under "How we make things," and burning gasoline is included in "How we get around." The same goes for things like cars, planes, and ships. The steel that they're made of is counted under "How we make things," and the emissions from the fuels they burn go under "How we get around."

Tip: Remember that emissions come from five different activities, and we need solutions in all of them.

3. How Much Power Are We Talking About?

This question mostly comes up in articles about electricity. You might read that some new power plant will produce 500 megawatts. Is that a lot? And what's a megawatt, anyway?

A megawatt is a million watts, and a watt is a joule per second. For our purposes, it doesn't matter what a joule is, other than a bit of energy. Just remember that a watt is a bit of energy per second. Think of it like this: If you were measuring the flow of water out of your kitchen faucet, you might count how many cups came out per second. Measuring power is similar, only you're measuring the flow of energy instead of water. Watts are equivalent to "cups per second."

A watt is pretty small. A small incandescent bulb might use 40 of them. A hair dryer uses 1,500. A power plant might generate hundreds of millions of watts. The largest power station in the world, the Three Gorges Dam in China, can produce 22 billion watts. (Remember that the definition of a watt already includes "per second," so there's no such thing as watts per second, or watts per hour. It's just watts.)

Because these numbers get big fast, it's convenient to use some shorthand. A kilowatt is 1,000 watts, a megawatt is a million, and a gigawatt (pronounced with a hard *g*!) is a billion. You often see this shorthand in the news, so I'll use it too.

The chart on the next page shows some rough comparisons that help me keep it all straight.

Of course, there's quite a bit of variation within these categories, throughout the day and throughout the year. Some homes use much more electricity than others. New York City runs on upwards of 12

How much power does it take?

The world	**5,000 gigawatts**
The United States	**1,000 gigawatts**
Mid-size city	**1 gigawatt**
Small town	**1 megawatt**
Average American house	**1 kilowatt**

gigawatts, depending on the season; Tokyo, with a larger population than New York, needs something like 23 gigawatts on average but can demand more than 50 gigawatts at peak use during the summer.

So let's say you want to power a mid-size city that requires a gigawatt. Could you simply build any one-gigawatt power station and guarantee that city all the electricity it'll need? Not necessarily. The answer depends on what your power source is, because some are more intermittent than others. A nuclear plant runs 24 hours a day and is shut down only for maintenance and refueling. But the wind doesn't always blow and the sun doesn't always shine, so the effective capacity of plants powered by wind and solar panels might be 30 percent or less. On average, they'll produce 30 percent of the gigawatt you need. That means you'll need to supplement them with other sources to get one gigawatt reliably.

Tip: Whenever you hear "kilowatt," think "house." "Gigawatt," think "city." A hundred or more gigawatts, think "big country."

4. How Much Space Do You Need?

Some power sources take up more room than others. This matters for the obvious reason that there is only so much land and water to go around. Space is far from the only consideration, of course, but it's an important one that we should be talking about more often than we do.

Power density is the relevant number here. It tells you how much power you can get from different sources for a given amount of land (or water, if you're putting wind turbines in the ocean). It's measured in watts per square meter. Below are a few examples:

How much power can we generate per square meter?

Energy source	Watts per square meter
Fossil fuels	500–10,000
Nuclear	500–1,000
Solar*	5–20
Hydropower (dams)	5–50
Wind	1–2
Wood and other biomass	Less than 1

* The power density of solar could theoretically reach 100 watts per square meter, though no one has accomplished this yet.

Notice that the power density of solar is considerably higher than that of wind. If you want to use wind instead of solar, you'll need far more land, all other things being equal. This doesn't mean that wind is bad and solar is good. It just means they have different requirements that should be part of the conversation.

Tip: If someone tells you that some source (wind, solar, nuclear, whatever) can supply all the energy the world needs, find out how much space will be required to produce that much energy.

5. How Much Is This Going to Cost?

The reason the world emits so much greenhouse gas is that—as long as you ignore the long-term damage they do—our current energy technologies are by and large the cheapest ones available. So moving our immense energy economy from "dirty," carbon-emitting technologies to ones with zero emissions will cost something.

How much? In some cases, we can price the difference directly. If we have a dirty source and a clean source of essentially the same thing, then we can just compare the price.

Most of these zero-carbon solutions are more expensive than their fossil-fuel counterparts. In part, that's because the prices of fossil fuels don't reflect the environmental damage they inflict, so they seem cheaper than the alternative. (I'll return to this challenge of pricing carbon in chapter 10.) These additional costs are what I call Green Premiums.*

During every conversation I have about climate change, Green Premiums are in the back of my mind. I'll come back to this concept repeatedly in the next several chapters, so I want to take a moment to explain what it means.

There isn't one single Green Premium. There are many: some for electricity, others for various fuels, others for cement, and so on. The size of the Green Premium depends on what you're replacing and what you're replacing it with. The cost of, say, zero-carbon jet fuel isn't the same as the cost of solar-generated electricity. I'll give you an example of how Green Premiums work in practice.

The average retail price for a gallon of jet fuel in the United States over the past few years is $2.22. Advanced biofuels for jets, to the extent they're available, cost on average $5.35 per gallon. The Green Premium for zero-carbon fuel, then, is the difference between these two prices, which is $3.13. That's a premium of more than 140 percent. (I'll explain all of this in more detail in chapter 7.)

In rare cases, a Green Premium can be negative; that is, going green can be *cheaper* than sticking with fossil fuels. For instance,

---

* I consulted with many people about the Green Premium, including experts at the Rhodium Group, Evolved Energy Research, and climate researcher Dr. Ken Caldeira. For information on how the Green Premiums in this book were calculated, visit breakthroughenergy.org.

depending on where you live, you may be able to save money by replacing your natural gas furnace and your air conditioner with an electric heat pump. In Oakland, doing this will save you 14 percent on your heating and cooling costs, while in Houston, the savings amount to 17 percent.

You might think that a technology with a negative Green Premium would already have been adopted around the world. By and large that is the case, but there is usually a lag between the introduction of a new technology and its being deployed—particularly for something like home furnaces, which we don't replace very often.

Once you've figured Green Premiums for all the big zero-carbon options, you can start having serious conversations about trade-offs. How much are we willing to pay to go green? Will we buy advanced biofuels that are twice as expensive as jet fuel? Will we buy green cement that costs twice as much as the conventional stuff?

By the way, when I ask, "What are we willing to pay?" I mean "we" in the global sense. It's not just a matter of what Americans and Europeans can afford. You can imagine Green Premiums high enough that the United States is willing and able to pay them but India, China, Nigeria, and Mexico are not. We need the premiums to be so low that everyone will be able to decarbonize.

Admittedly, Green Premiums are a moving target. A lot of assumptions go into estimating them; for this book, I've made the assumptions that seem reasonable to me, but different well-informed people would make different assumptions and arrive at different numbers. What's more important than the specific prices is knowing whether a given green technology is close to being as cheap as its fossil-fuel counterpart and, for the ones that aren't close, thinking about how innovation might bring their prices down.

I hope the Green Premiums in this book will be the start of a longer conversation about the costs of getting to zero. I hope other people will do their own calculations of the premiums, and I'd be especially happy to learn that some of them aren't as high as I think.

The ones I've calculated in this book are an imperfect tool for comparing costs, but they're better than no tool at all.

In particular, Green Premiums are a fantastic lens for making decisions. They help us put our time, attention, and money to their best use. Looking at all the different premiums, we can decide which zero-carbon solutions we should deploy now and where we should pursue breakthroughs because the clean alternatives aren't cheap enough. They help us answer questions like these:

Which zero-carbon options should we be deploying now?
Answer: the ones with a low Green Premium, or no premium at all. If we're not deploying these solutions already, it's a sign that cost isn't the barrier. Something else—like outdated public policies or lack of awareness—is stopping us from getting them out there in a big way.

Where do we need to focus our research and development spending, our early investors, and our best inventors?
Answer: wherever we decide Green Premiums are too high. That's where the extra cost of going green will keep us from decarbonizing and where there's an opening for new technologies, companies, and products that make it affordable. Countries that excel at research and development can create new products, make them more affordable, and export them to the places that can't pay the current premiums. Then no one will have to argue about whether every nation is doing its fair share to avoid a climate disaster; instead, countries and companies will be racing to create and market the affordable innovations that help the world get to zero.

There's one last benefit to the Green Premium concept: It can act as a measurement system that shows us the progress we're making toward stopping climate change.

In that sense, Green Premiums remind me of a problem that Melinda and I encountered when we first started working in global health. Experts could tell us how many children died around the world every year, but they couldn't tell us much about what caused those deaths. We knew that a certain number of kids died of diarrhea, but we didn't know what caused the diarrhea in the first place. How could we know which innovations might save lives if we didn't know why children were dying?

So working with partners around the world, we funded various studies to find out what was killing children. Eventually, we were able to track deaths with much more detail, and this data pointed the way to big breakthroughs. For example, we saw that pneumonia was behind a large number of children's deaths each year. Although a pneumo vaccine already existed, it was so expensive that poor countries weren't buying it. (They had little incentive to, because they had no idea how many children were dying from the disease.) Once they saw the data, though—and once donors agreed to pay most of the cost—they began adding the vaccine to their health programs, and eventually we were able to fund a much cheaper vaccine that's now in use in countries around the world.

The Green Premiums can do something similar for greenhouse gas emissions. The premiums give us a different insight from the raw number of emissions, which shows us how far we are from zero but tells us nothing about how hard it will be *to get there*. What would it cost to use the zero-carbon tools we have now? Which innovations will make the biggest impact on emissions? The Green Premiums answer these questions, measuring the cost of getting to zero, sector by sector, and highlighting where we need to innovate—just as the data showed us that we needed to make a big push for the pneumo vaccine.

In some cases, such as the jet fuel example I mentioned earlier, the direct approach to estimating Green Premiums is simple. But when we apply it more generally, we have a problem: We don't currently have a direct green equivalent in every case. There's no such thing as zero-carbon cement (at least not yet). How do we get a sense of the cost of a green solution in those cases?

We can do it by conducting a thought experiment. "How much would it cost to just suck the carbon out of the atmosphere directly?" That idea has a name; it's called direct air capture. (In short, with DAC you blow air over a device that absorbs carbon dioxide, and then you store the gas for safekeeping.) DAC is an expensive and largely unproven technology, but if it can work at a large scale, it would allow us to capture carbon dioxide no matter when or where it was produced. The one DAC facility now in operation, which is based in Switzerland, is absorbing gas that might have been spewed out by a coal-fired plant in Texas 10 years ago.

To figure out how much this approach would cost, we need just two data points: the amount of global emissions, and the cost of absorbing emissions using DAC.

We already know the emissions number; it's 52 billion tons each year. As for the cost of removing a ton of carbon from the air, that figure hasn't been firmly established, but it's at least $200 per ton. With some innovation, I think we can realistically expect it to get down to $100 per ton, so that's the number I'll use.

That gives us the following equation:

52 billion tons per year x $100 per ton = $5.2 trillion per year

In other words, using the DAC approach to solve the climate problem would cost at least $5.2 trillion per year, every year, as long as we produce emissions. That's around 6 percent of the world's economy. (It's an enormous sum, though this theoretical DAC technology would actually be far cheaper than the cost of trying to reduce emissions by shutting down sectors of the economy, as we've done during the COVID-19 pandemic. In the United States, according to data from the Rhodium Group, the per-ton cost to our economy came to between $2,600 and $3,300. In the European Union, it was more than $4,000 per ton. In other words, it cost between 25 and 40 times the $100 per ton we hope to achieve someday.)

As I mentioned, the DAC-based approach is really just a thought experiment. In reality, the technology behind DAC isn't ready for global deployment, and even if it were, DAC would be an extremely inefficient method for solving the world's carbon problem. It's not clear that we could store hundreds of billions of tons of carbon safely. There's no practical way to collect $5.2 trillion a year or make sure everyone pays their fair share (and even defining everyone's fair share would be a major political fight). We'd need to build more than 50,000 DAC plants around the world just to manage the emissions we're producing right now. In addition, DAC doesn't work on methane or other greenhouse gases, just carbon dioxide. And it's probably the most expensive solution; in many cases, it will be cheaper not to emit greenhouse gases in the first place.

Even if DAC can eventually be made to work on a global scale—and remember that I'm an optimist when it comes to technology—it almost certainly can't be developed and deployed quickly enough to prevent dire harm to the environment. Unfortunately, we can't just wait for a future technology like DAC to save us. We have to start saving ourselves today.

Tip: Keep the Green Premiums in mind and ask whether they're low enough for middle-income countries to pay.

Here's a summary of all five tips:

1. Convert tons of emissions to a percentage of 52 billion.
2. Remember that we need to find solutions for all five activities that emissions come from: making things, plugging in, growing things, getting around, and keeping cool and warm.

3. Kilowatt = house. Gigawatt = mid-size city. Hundreds of gigawatts = big, rich country.
4. Consider how much space you're going to need.
5. Keep the Green Premiums in mind and ask whether they're low enough for middle-income countries to pay.

CHAPTER 4

HOW WE PLUG IN

26 percent of 52 billion tons per year

We're in love with electricity, but most of us don't know it. Electricity is consistently there for us, making sure our streetlights, air conditioners, computers, and TVs always work. It powers all sorts of industrial processes most of us would rather not think about. But, as sometimes happens in life, we don't realize how much it means to us until it's gone. In the United States, power outages are so rare that people remember that one time a decade ago when the lights went out and they got stuck in an elevator.

I wasn't always aware of how much we rely on electricity, but over the years I've gradually come to see how essential it is. And I really appreciate what it takes to deliver this miracle. In fact, it's fair to say that I'm in awe of all the physical infrastructure that makes electricity so cheap, available, and reliable. It's downright magical that you can simply turn a switch almost anywhere in a well-off country and expect the lights to come on for a fraction of a penny. Literally: In the United States, leaving a 40-watt lightbulb turned on for an hour costs you about half of one cent.

I'm not the only one in the family who feels this way about electricity: My son, Rory, and I used to visit power plants for fun, just to learn how they worked.

After a family visit to the Þríhnúkagígur volcano in Iceland in 2015,
Rory and I checked out the geothermal power plant next door.

I'm glad I've invested all that time learning about electricity. For
one thing, it was a great father-son activity. (Seriously.) Besides,
figuring out how to get all the benefits of cheap, reliable electricity without emitting greenhouse gases is the single most important thing we must do to avoid a climate disaster. That's partly
because producing electricity is a major contributor to climate
change, and also because, if we get zero-carbon electricity, we can
use it to help decarbonize lots of other activities, like how we get
around and how we make things. The energy we give up by not
using coal, natural gas, and oil has to come from somewhere, and
mostly it will come from clean electricity. This is why I'm covering
electricity first, even though manufacturing is responsible for more
emissions.

Plus, even *more* people should be getting and using electricity.
In sub-Saharan Africa, less than half of the population has reliable

power at home.* And if you don't have access to any electricity at all, even a seemingly simple task like recharging your mobile phone is difficult and expensive. You have to walk to a store and pay 25 cents or more to plug your phone into an outlet, hundreds of times more than people pay in developed countries.

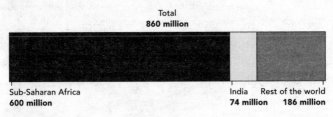

Total
860 million

Sub-Saharan Africa
600 million

India
74 million

Rest of the world
186 million

860 million people don't have reliable access to electricity. Fewer than half the people in sub-Saharan Africa are on the grid. (IEA)

I don't expect most people to get as excited about grids and transformers as I do. (Even I can recognize that you have to be a pretty big nerd to write a sentence like "I'm in awe of physical infrastructure.") But I think if everyone stopped to consider what it takes to deliver the service we now take for granted, they would appreciate it more. And they'd realize that none of us want to give it up. Whatever methods we use to get to zero-carbon electricity in the future will have to be as dependable and nearly as affordable as the ones we use today.

In this chapter I want to explain what it will take to keep getting all the things we like from electricity—a cheap source of energy that's always available—and deliver it to even more people, but without the carbon emissions. The story starts with how we got here and where we're headed.

---

* I'm using the word "power" a bit loosely here. Technically, "power" refers to the rate of flow of electricity, measured in watts. In this book, for the sake of readability, I'll use the term in its more general sense, as a synonym for "electricity."

—

Considering how ubiquitous electricity is today, it's easy to forget that it only became an important factor in most Americans' lives a few decades into the 20th century. And one of our early major sources of electricity wasn't any of the ones that we think of today, like coal, oil, or natural gas. It was water, in the form of hydropower.

Hydropower has a lot going for it—it's relatively cheap—but it also has some big downsides. Making a reservoir displaces local communities and wildlife. When you cover land with water, if there's a lot of carbon in the soil, the carbon eventually turns into methane and escapes into the atmosphere—which is why studies show that depending on where it's built, a dam can actually be a worse emitter than coal for 50 to 100 years before it makes up for all the methane it's responsible for.* In addition, the amount of electricity you can generate from a dam depends on the season, because you're relying on rain-fed streams and rivers. And, of course, hydropower is immobile. You have to build the dams where the rivers are.

Fossil fuels don't have that limitation. You can take coal, oil, or natural gas out of the ground and move it to a power plant, where you burn it, use the heat to boil water, and let the steam from the boiling water turn a turbine to make electricity.

Because of all these advantages, when demand for electricity in the United States took off after World War II, we met it with fossil fuels. They provided most of the new capacity we built in the second

* These calculations are drawn from a life cycle assessment of dams. Life cycle assessment is an interesting field that involves documenting all the greenhouse gases that a given product is responsible for, from the time it's produced until the end of its life. These assessments are a useful way to analyze the climate impact of various technologies, but they're pretty complicated, so in this book I will focus on direct emissions, which are easier to explain and generally lead to the same conclusions anyway.

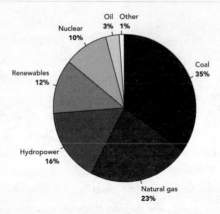

Getting all the world's electricity from clean sources won't be easy. Today, fossil fuels account for two-thirds of all electricity generated worldwide. (bp Statistical Review of World Energy 2021)

half of the 20th century—some 700 gigawatts, nearly 60 times more than we had installed before the war.

Over time, electricity has become extraordinarily cheap. One study found that it was at least 200 times more affordable in the year 2000 than in 1900. Today, the United States spends only 2 percent of its GDP on electricity, an amazingly low number when you consider how much we rely on it.

The main reason it's so cheap is that fossil fuels are cheap. They're widely available, and we've developed better and more efficient ways to extract them and turn them into electricity. Governments also go to considerable effort to keep the prices of fossil fuels low and encourage their production.

In the United States, we've been doing this since the earliest days of the Republic: Congress enacted America's first protective tariff on imported coal in 1789. In the early 1800s, recognizing how important coal was for the railroad industry, states began to exempt it from some taxes and created other incentives for its production. After the corporate income tax was established in 1913, oil and gas producers got the right to deduct certain expenses, including drilling costs. In all, these tax expenditures represented roughly $42 billion (in today's dollars) in support for coal and natural gas producers from 1950 through 1978, and they're still in the tax code today. In addition, coal and gas producers benefit from favorable leasing terms on federal lands.

This flyer featuring a coal facility in Connellsville, Pennsylvania, dates from around 1900.

The United States isn't alone. Most countries take various steps to keep fossil fuels cheap—the International Energy Agency (IEA) estimates that government subsidies for the consumption of fossil fuels amounted to $400 billion in 2018—which helps explain why they're such a steady part of our electricity supply. The share of global power that comes from burning coal (roughly 40 percent) hasn't changed in 30 years. Oil and natural gas together have been hovering around 26 percent for three decades. All told, fossil fuels provide two-thirds of the world's electricity. Solar and wind, meanwhile, account for 9 percent.

As of mid-2019, some 236 gigawatts' worth of coal plants were being built around the world; coal and natural gas are now the fuels of choice in developing countries, where demand has skyrocketed in the past few decades. Between 2000 and 2018, China tripled the amount of coal power it uses. That's more capacity than in the United States, Mexico, and Canada combined!

Can we turn this around and get all the electricity we'll need without any greenhouse gas emissions?

It depends on what you mean by "we." The United States can get pretty close, with the right policies to expand wind and solar along with a big push for specific innovations. But can the whole world get zero-carbon electricity? That will be much harder.

Let's start with the Green Premiums for electricity in the United States. It's actually good news: We can eliminate our emissions with only a modest Green Premium.

In the case of electricity, the premium is the additional cost of getting all our power from non-emitting sources, including wind, solar, nuclear power, and coal- and natural-gas-fired plants equipped with devices that capture the carbon they produce. (Remember that the goal isn't to use only renewable sources like wind and solar; the goal is to get to zero emissions. That's why I'm including these other zero-carbon options.)

How much is the premium? Changing America's entire electricity system to zero-carbon sources would raise average retail rates by between 1.3 and 1.7 cents per kilowatt-hour, roughly 15 percent more than what most people pay now. That adds up to a Green Premium of $18 a month for the average home—pretty affordable for most people, though possibly not for low-income Americans, who already spend a tenth of their income on energy.

(You're probably familiar with kilowatt-hours if you pay a utility bill, because they're how we're charged for electricity in our homes. But in case you're wondering, a kilowatt-hour is a unit of energy that's used to measure how much electricity you use in a given time period. If you consume one kilowatt for an hour, you've used one kilowatt-hour. The typical U.S. household uses 29 kilowatt-hours a day. On average, across all types of customers and states in the United States, a kilowatt-hour of electricity costs around 10 cents, though in some places it can be more than three times that much.)

It's great that America's Green Premium could be so low. Europe is similarly well situated; one study by a European trade association suggested that decarbonizing its power grid by 90 to 95 percent would cause average rates to go up about 20 percent. (This study used a different methodology from the way I figured America's Green Premium.)

Unfortunately, few other countries are so lucky. The United States has a large supply of renewables, including hydropower in the Pacific Northwest, strong winds in the Midwest, and year-round solar power in the Southwest and California. Other countries might have some sun but no wind, or some wind but little year-round sun, or not much of either. And they might have low credit ratings that make it hard to finance big investments in new power plants.

Africa and Asia are in the toughest position. Over the past few decades, China has accomplished one of the greatest feats in history—lifting hundreds of millions of people out of poverty—and did it in part by building coal-fired electric plants very cheaply. Chinese firms drove down the cost of a coal plant by a remarkable 75 percent. And now they understandably want more customers, so they're making a big play to attract the next wave of developing countries: India, Indonesia, Vietnam, Pakistan, and nations throughout Africa.

What will those potential new customers do? Will they build

coal plants or go clean? Consider their goals and their options. Small-scale solar can be an option for people in poor, rural areas who need to charge their cell phones and run lights at night. But that kind of solution is never going to deliver the massive amounts of cheap, always-available electricity these countries need to jump-start their economies. They're looking to do what China did: grow their economies by attracting industries like manufacturing and call centers—the types of businesses that demand far more (and far more reliable) power than small-scale renewables can provide today.

If these countries opt for coal plants, as China and every rich country did, it'll be a disaster for the climate. But right now, that's their most economical option.

It's not immediately obvious why there's such a thing as a Green Premium in the first place. Natural gas plants have to keep buying fuel as long as they're running; solar farms, wind farms, and dams get their fuel for free. Also, there's the truism that as you take a technology to broad scale, it gets cheaper. So why does it cost extra to go green?

One problem is that fossil fuels are so cheap. Because their prices don't factor in the true cost of climate change—the economic damage they inflict by making the planet warmer—it's harder for clean energy sources to compete with them. And we've spent many decades building up a system to extract fossil fuels from the ground, get energy from them, and deliver that energy, all very cheaply.

Another reason is that, as I mentioned earlier, some regions of the world simply don't have decent renewable resources. To get close to 100 percent, we'd have to move lots of clean energy from where it's made (sunny places, ideally near the equator, and windy regions) to where it's needed (cloudy, windless ones). That would require building new transmission lines, a costly and time-consuming task—especially if it involves crossing national borders—and the more power lines we install, the more the price of power goes up. In fact, transmission and distribution are responsible for more than a third of the final cost of electricity.* And many countries don't want to rely on other countries for their electricity supply.

But cheap oil and expensive transmission lines aren't the biggest drivers of the electricity Green Premium. The main culprits are our demand for reliability, and the curse of intermittency.

The sun and the wind are intermittent sources, meaning that they don't generate electricity 24 hours a day, 365 days a year. But our need for power is not intermittent; we want it all the time. So if solar and wind represent a big part of our electricity mix and we want to avoid major outages, we're going to need other options for when the sun isn't shining and the wind isn't blowing. Either we need to store excess electricity in batteries (which, I'll argue in a moment, is prohibitively expensive), or we need to add other energy sources that use fossil fuels, such as natural gas plants that run only when you need them. Either way, the economics won't work in our favor. As we approach 100 percent clean electricity, intermittency becomes a bigger and more expensive problem.

The clearest example of intermittency is when the sun goes down, cutting off our supply of solar-generated electricity. Suppose we want to solve that problem by taking one kilowatt-hour of excess electricity that's generated during the day, storing it, and using it that night. (You'd need much more than that, but I'm picking one kilowatt-hour to make the math easy.) How much would that add to our electric bill?

---

* Think of transmission as the freeway and distribution as the local road. We use high-voltage transmission lines to deliver electricity from the power plant to the city. Then the electricity goes into the local, lower-voltage distribution system—the power lines you see in your neighborhood.

That depends on two factors: how much the battery costs, and how long it'll last before we have to replace it. For the cost, let's say we can buy a one-kilowatt-hour battery for $100. (This is a conservative estimate, and I'll ignore for the moment what happens if we have to take out a loan for this battery.) As for how long our battery will last, let's assume it can go through 1,000 charge-and-discharge cycles.

So the capital cost of this one-kilowatt-hour battery will be $100 spread out over 1,000 cycles, which works out to 10 cents per kilowatt-hour. That's on top of the cost of generating the power in the first place, which in the case of solar power is something like 5 cents per kilowatt-hour. In other words, the electricity we store for nighttime use will cost us *triple* what we're paying during the day—5 cents to generate and 10 cents to store, for a total of 15 cents.

I know researchers who think they can make a battery that lasts five times longer than the one I've described here. They haven't done it yet, but if they're right, that would drive the premium down from 10 cents to 2 cents, a much more modest increase. In any case, the nighttime problem is solvable today, if you're willing to pay a big premium, and with innovation I'm confident we can reduce that premium.

Unfortunately, nighttime intermittency isn't the hardest problem to deal with. The seasonal variation between summer and winter is an even bigger hurdle. There are various ways to try to deal with it—like adding in power from a nuclear plant or a gas-fired electric plant fitted with a device that captures its emissions—and any realistic scenario will include these options. I'll get to them later in the chapter, but for the sake of simplicity for now I'll just use batteries to illustrate the problem of seasonal variation.

Say we want to store a single kilowatt-hour not for a day but for a whole season. We'll generate it during the summer and use it in the winter to run a space heater. This time, the battery's life cycle isn't an issue, because we're charging it only once a year.

But suppose we have to finance the purchase of the battery. Now we've tied up $100 in capital. (Obviously you wouldn't finance a $100 battery, but you might need financing if you were buying enough to store several gigawatts. And the math is the same.) If we pay 5 percent interest on the capital, and the battery costs $100, that's an additional $5 cost to store our single kilowatt-hour. And remember how much we're paying for solar power during the day: just 5 cents. Who would pay $5 to store a nickel's worth of electricity?

Seasonal intermittency and the high cost of storage cause yet another problem, especially for big users of solar power—the problem of overgeneration in the summer and undergeneration in the winter.

Because the earth is tilted on its axis, the amount of sunlight that hits any given part of the planet varies across the four seasons, as does the intensity of the sunlight. Just how big the variation is depends on how far you are from the equator. In Ecuador, there's essentially no change. In the Seattle area, where I live, we get about twice as much sunlight on the longest day of the year as on the shortest day. Parts of Canada and Russia get about 12 times more.*

To see why this variation matters, let's do another thought experiment. Imagine there's a town near Seattle—we'll call it Suntown—that wants to generate a gigawatt of solar power year-round. How big should Suntown's solar array be?

One option would be to install enough panels to produce a gigawatt during the summer, when sunlight is plentiful. But the town would be out of luck in the winter, when it'll get only half as much sunlight. That's undergeneration. (And the town council is well aware that storage is excessively expensive, so they've ruled

---

* There's also seasonal variation for wind. In the United States, wind power tends to be at its peak in the spring and reach its low point in mid- to late summer (although it's the opposite in California). The difference can be a multiple of two to four.

out batteries.) On the other hand, Suntown could put up all the solar panels it needs for the short, dark days of winter, but then by the time summer arrives, it would be generating way more than necessary. Electricity would be so cheap that the town would be hard-pressed to recoup the expense of installing all those panels.

Suntown could deal with this overgeneration problem by turning off some of its panels during the summer, but then it'd be sinking money into equipment that gets used only for part of the year. That would raise the cost of electricity even more for every home and business in town; in other words, it would add to the town's Green Premium.

The situation with Suntown isn't merely a hypothetical example. Something similar has been happening in Germany, which through its ambitious Energiewende program set a goal of 60 percent renewables by 2050. The country has spent billions of dollars over the past decade expanding its use of renewables, increasing its solar capacity nearly 650 percent between 2008 and 2010. But Germany produced about 10 times more solar in June 2018 than it did in December 2018. In fact, at times during the summer, Germany's solar and wind plants generate so much electricity that the country can't use it all. When that happens, it ends up transmitting some of the excess to neighboring Poland and the Czech Republic, whose leaders have complained that it's straining their own power grids and causing unpredictable swings in the cost of electricity.

There's another problem caused by intermittency, and it's even harder to solve than the daily or seasonal variety. What happens when an extreme event forces a city to survive several days without any renewable energy at all?

Imagine a future where Tokyo gets all its electricity from wind power alone. (Japan does, in fact, have quite a bit of onshore and offshore wind available.) One August, at the peak of cyclone season, a massive storm hits. The winds are so strong that they will rip the city's wind turbines apart if they aren't shut down. Tokyo's leaders decide to switch off the turbines and get by solely on electricity stored up in the best large-scale batteries they can find.

Here's the question: How many batteries would they need in order to power Tokyo for three days, until the storm passes and they can turn the turbines back on?

The answer is more than 14 million batteries. That's more storage capacity than the world produces in seven years. Purchase price: $400 billion. Averaged over the lifetime of the batteries, that's an annual expense of more than $27 billion.* And that's just the capital cost of the batteries; it doesn't include other expenses like installation and maintenance.

This example is entirely hypothetical. No one seriously thinks Tokyo should get all its electricity from wind or store all of it in today's batteries. I'm using this illustration to make a crucial point: It's extremely difficult and expensive to store electricity on a large scale, but that's one of the things we'll need to do if we're going to rely on intermittent sources to provide a significant percentage of clean electricity in the coming years.

And we're going to need *much* more clean electricity in the coming years. Most experts agree that as we electrify other carbon-intensive processes like making steel and running cars, the world's electricity supply will need to double or even triple by 2050. And that doesn't even account for population growth, or the fact that people will get richer and use more electricity. So the world will need much more than three times the electricity we generate now.

Because solar and wind are intermittent, our *capacity* to generate electricity will need to grow even more. (Capacity measures how much electricity we're theoretically capable of producing when the

---

* Here's how I got these figures: Between August 6 and August 8, 2019, Tokyo consumed 3,122 gigawatt-hours of electricity. For baseload power, I assumed 5.4 million iron-flow batteries with a system lifetime of 20 years and a per-unit cost of $36,000. For peak demand, I assumed 9.1 million lithium-ion batteries with a system lifetime of 10 years and a per-unit cost of $23,300.

sun is shining its brightest or the wind is blowing its hardest; generation is how much we actually get, after accounting for intermittency, shutting down power plants for maintenance, and other factors. Generation is always smaller than capacity, and in the case of variable sources like solar and wind it can be a lot smaller.)

With all the additional electricity we'll be using, and assuming that wind and solar play a significant role, completely decarbonizing America's power grid by 2050 will require adding around 75 gigawatts of capacity every year for the next 30 years.

Is that a lot? Over the past decade, we've added an average of 22 gigawatts a year. Now we need to install more than three times that much each year, and keep up the pace for the next three decades.

That will be a bit easier as we make solar panels and wind turbines cheaper and even more efficient—that is, as we invent ways to get even more energy from a given amount of sunlight or wind. (The best solar panels today convert less than a quarter of the sunlight that hits them into electricity, and the theoretical limit for the most common type of commercially available panels is about 33 percent.) As these conversion rates go up, we can get more power from the same amount of land, which will help as we deploy these technologies widely.

But more efficient panels and turbines aren't enough, because there's a major difference between the build-out America did in the 20th century and what we need to do in the 21st. Location is going to matter more than ever.

Since the beginning of the electric grid, utilities have placed most power plants close to America's rapidly growing cities, because it was relatively easy to use railroads and pipelines to ship fossil fuels from wherever they were extracted to the power plants where they'd be burned to make electricity. As a result, America's power grid relies on railroads and pipelines to move fuels over long distances to power plants, and then on transmission lines to move electricity over short distances to the cities that need it.

That model doesn't work with solar and wind. You can't ship sunlight in a railcar to some power plant; it has to be converted to electricity on the spot. But most of America's sunlight supply is in the Southwest, and most of our wind is in the Great Plains, far from many major urban areas.

In short, intermittency is the main force that pushes the cost up as we get closer to all zero-carbon electricity. It's why cities that are trying to go green still supplement solar and wind with other ways to generate electricity, such as gas-fired power plants that can be powered up and down as needed to meet demand, and these so-called peakers are not zero-carbon by any stretch of the imagination.

Just to be clear: Variable energy sources like solar and wind can play a substantial role in getting us to zero. In fact, *we need them to*. We should be deploying renewables quickly wherever it's economical to do so. It's amazing how much the costs of solar and wind power have dropped in the past decade: Solar cells, for example, got almost 10 times cheaper between 2010 and 2020, and the price of a full solar system went down by 11 percent in 2019 alone. A lot of the credit for these decreases goes to learning by doing—the simple fact that the more times we make some product, the better we get at it.

We do need to remove the barriers that keep us from making the most of renewable sources. For example, it's natural to think of America's electric grid as one single connected network, but in reality it's nothing of the sort. There isn't one power grid; there are many, and they're a patchwork mess that makes it essentially impossible to send electricity beyond the region where it's made. Arizona can sell spare solar power to its neighbors, but not to a state on the other side of the country.

We could solve this problem by crisscrossing the country with thousands of miles of special long-distance power lines carrying what's called high-voltage current. This technology already exists; in fact, the United States already has some of these lines installed. (The biggest one runs from Washington State to California.) But the political hurdles to a massive upgrade of our electric grid are considerable.

Just think about how many landowners, utility companies, and local and state governments you'd need to bring together to build power lines that could move solar energy from the Southwest all the way to customers in New England. Merely picking the routes and establishing rights-of-way would be a massive undertaking; people tend to object when you want to run a big power line through the local park.

Construction on the TransWest Express, a transmission project designed to move wind-generated power from Wyoming to California and the Southwest, is scheduled to begin in 2021. The project is supposed to become operational in 2024—some 17 years after planning began.

But if we could pull this off, it would be transformative. I'm funding a project that involves building a computer model of all the power grids covering the United States. Using the model, experts have studied what it would take for all western states to reach California's goal of 60 percent renewables by 2030, and for all eastern states to reach New York's goal of 70 percent clean energy by that same year. What they found is that there's simply no way for the states to do it without enhancing the power grid. The model also showed that regional and national approaches to transmission—rather than leaving each state to its own devices—would allow every state to meet the emission-reduction goals with 30 percent fewer renewables than they would need otherwise. In other words, we'll save money by building renewables in the best locations, building a unified national grid, and shipping zero-emissions electrons wherever they're needed.*

In the coming years, as electricity becomes an even bigger part of

---

* This model is available online for the public. See breakthroughenergy.org for more information.

our overall energy diet, we'll need models like these for grids around the world. They'll help us answer questions like: Which mix of clean energy sources will be the most efficient in a given place? Where should transmission lines go? Which regulations stand in the way, and what incentives do we need to create? I hope to see a lot more projects like this one.

Here's another complication: As our houses rely less on fossil fuels and more on electricity (for example, to power electric cars and stay warm in the winter), we'll need to upgrade the electrical service to each household—by at least a factor of two, and in many cases even more than that. A lot of streets will need to be dug up and electrical poles climbed to install heavier wires, transformers, and other equipment. So it will be felt in a real way by nearly every community, and the political impact will get down to the local level.

Technology might be able to help overcome some of the political barriers involved with these upgrades. For example, power lines are less of an eyesore if they're run underground. But today, burying power lines increases the cost by a factor of 5 to 10. (The problem is heat: Power lines get hot when there's electricity running through them. That's no problem when they're aboveground—the heat just dissipates into the air—but underground there's no place for the heat to go. If the temperature gets too high, the power lines melt.) Some companies are working on next-generation transmission that would eliminate the heat problem and reduce the cost of underground lines significantly.

Deploying today's renewables and improving transmission couldn't be more important. If we don't upgrade our grid significantly and instead make each region do this on its own, the Green Premium might not be 15 to 30 percent; it could be 100 percent or more. Unless we use large amounts of nuclear energy (which I'll get to in the next section), every path to zero in the United States will require us to install as much wind and solar power as we can build and find room for. It's hard to say exactly how much of America's electricity will come from renewables in the end, but what we do know is that between now and 2050 we have to build them much faster—on the order of 5 to 10 times faster—than we're doing right now.

And remember that most countries aren't as lucky as the United States when it comes to solar and wind resources. The fact that we can hope to generate a large percentage of our power from renewables is the exception rather than the rule. That's why, even as we deploy, deploy, deploy solar and wind, the world is going to need some new clean electricity inventions too.

There's already a lot of great research going on. If there's one thing I love about my work, it's the opportunity to meet with, and learn from, top scientists and entrepreneurs. Over the years, through my investments in Breakthrough Energy and in other ways, I've heard about some potential breakthroughs that could be the revolution we need to get to zero emissions in electricity. These ideas are in various stages of development; some are relatively mature and well tested, while others are, frankly, nuts. But we can't be afraid to bet on some crazy ideas. It's the only way to guarantee at least a few breakthroughs.

Making Carbon-Free Electricity

Nuclear fission. Here's the one-sentence case for nuclear power: It's the only carbon-free energy source that can reliably deliver power day and night, through every season, almost anywhere on earth, that has been proven to work on a large scale.

No other clean energy source even comes close to what nuclear already provides today. (Here I mean nuclear fission—the process of getting energy by splitting atoms apart. I'll get to its counterpart, nuclear fusion, in the next section.) The United States gets around

20 percent of its electricity from nuclear plants; France has the highest share in the world, getting 70 percent of its electricity from nuclear. Remember that by comparison solar and wind together provide about 9 percent worldwide.

And it's hard to foresee a future where we decarbonize our power grid affordably without using more nuclear power. In 2018, researchers at the Massachusetts Institute of Technology analyzed nearly 1,000 scenarios for getting to zero in the United States; all the cheapest paths involved using a power source that's clean and always available—that is, one like nuclear power. Without a source like that, getting to zero-carbon electricity would cost a lot more.

Nuclear plants are also number one when it comes to efficiently using materials like cement, steel, and glass. This chart shows you how much material it takes to generate a unit of electricity from various sources:

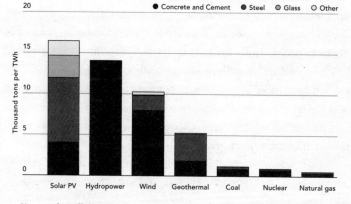

How much stuff does it take to build and run a power plant? That depends on the type of plant. Nuclear is the most efficient, using much less material per unit of electricity generated than other sources do. (U.S. Department of Energy)

See how small the nuclear stack is? That means you're getting far more energy for each pound of material that goes into building and running the power plant. It's a major consideration, given all the greenhouse gases that are emitted when we produce those materials.

(See the next chapter for more detail about that.) And these figures don't take into account the fact that solar and wind farms generally need more land than nuclear plants, and they generate power only 25 to 40 percent of the time, versus 90 percent for nuclear. So the difference is even more dramatic than this chart shows.

It's no secret that nuclear power has its problems. It's very expensive to build today. Human error can cause accidents. Uranium, the fuel it uses, can be converted for use in weapons. The waste is dangerous and hard to store.

High-profile accidents at Three Mile Island in the United States, Chernobyl in the former U.S.S.R., and Fukushima in Japan put a spotlight on all these risks. There are real problems that led to those disasters, but instead of getting to work on solving those problems, we just stopped trying to advance the field.

Imagine if everyone had gotten together one day and said, "Hey, cars are killing people. They're dangerous. Let's stop driving and give up these automobiles." That would've been ridiculous, of course. We did just the opposite: We used innovation to make cars safer. To keep people from flying through the windshield, we invented seat belts and air bags. To protect passengers during an accident, we created safer materials and better designs. To protect pedestrians in parking lots, we started installing rear-view cameras.

Nuclear power kills far, far fewer people than cars do. For that matter, it kills far fewer people than any fossil fuel.

Nevertheless, we should improve it, just as we did with cars, by analyzing the problems one by one and setting out to solve them with innovation.

Scientists and engineers have proposed various solutions. I'm very optimistic about the approach created by TerraPower, a company I founded in 2008, bringing together some of the best minds in nuclear physics and computer modeling to design a next-generation nuclear reactor.

Because no one was going to let us build experimental reactors

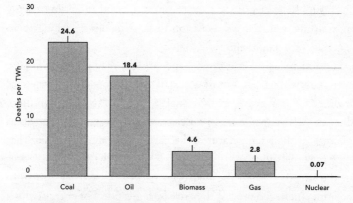

Is nuclear power dangerous? Not if you're counting the number of deaths caused per unit of electricity, as this chart shows. The numbers here cover the entire process of generating energy, from extracting fuels to turning them into electricity, as well as the environmental problems they cause, such as air pollution. (Our World in Data)

in the real world, we set up a lab of supercomputers in Bellevue, Washington, where the team runs digital simulations of different reactor designs. We think we've created a model that solves all the key problems using a design called a traveling wave reactor.

TerraPower's reactor could run on many different types of fuel, including the waste from other nuclear facilities. The reactor would produce far less waste than today's plants, would be fully automated—eliminating the possibility of human error—and could be built underground, protecting it from attack. Finally, the design would be inherently safe, using some ingenious features to control the nuclear reaction; for example, the radioactive fuel is contained in pins that expand if they get too hot, which slows the nuclear reaction down and prevents overheating. Accidents would literally be prevented by the laws of physics.

We're still years away from breaking ground on a new plant. So far, TerraPower's design exists only in our supercomputers; we're working with the U.S. government on building our first prototype.

Nuclear fusion. There's another, entirely different approach to nuclear power that's quite promising but still at least a decade away from supplying electricity to consumers. Instead of getting energy by splitting atoms apart, as fission does, it involves pushing them together, or fusing them.

Fusion relies on the same basic process that powers the sun. You start with a gas—most research focuses on certain types of hydrogen—and get it extraordinarily hot, well over 50 million degrees Celsius, while it's in an electrically charged state known as plasma. At these temperatures, the particles are moving so fast that they hit each other and fuse together, just as the hydrogen atoms in the sun do. When the hydrogen particles fuse, they turn into helium, and in the process they release a great deal of energy, which can be used to generate electricity. (Scientists have various ways of containing the plasma; the most common methods use either powerful magnets or lasers.)

Although it's still in the experimental phase, fusion holds a lot of promise. Because it would run on commonly available elements like hydrogen, the fuel would be cheap and plentiful. The main type of hydrogen that's usually used in fusion can be extracted from seawater, and there's enough of it to meet the world's energy needs for many thousands of years. Fusion's waste products would be radioactive for hundreds of years, versus hundreds of thousands for waste plutonium and other elements from fission, and at a much lower level—about as dangerous as radioactive hospital waste. There's no chain reaction to run out of control, because the fusion ceases as soon as you stop supplying fuel or switch off the device that's containing the plasma.

In practice, though, fusion is very hard to do. There's an old joke among nuclear scientists: "Fusion is 40 years away, and it always will be." (Admittedly, I'm using the term "joke" loosely.) One of the big hurdles is that it takes so much energy to kick off the fusion reaction that you often end up putting more into the process than you get out of it. And, as you might imagine given the temperatures involved, it's also a huge engineering challenge to build a reactor. None of the existing fusion reactors are designed to produce electricity that consumers could use; they're for research purposes only.

The biggest project currently under construction, a collaboration between six countries and the European Union, is an experimental facility in southern France known as ITER (pronounced like "eater"). Construction on the project began in 2010 and is still ongoing. By the mid-2020s, ITER is expected to generate its first plasma, and to generate excess power—10 times more than it needs to operate—in the late 2030s. That would be the Kitty Hawk moment for fusion, a major accomplishment that would put us on the path to building a commercial demonstration plant.

And there are more innovations coming that could make fusion more practical. For example, I know of companies that are using high-temperature superconductors to make much stronger magnetic fields for containing the plasma. If this approach works, it would allow us to make fusion reactors far smaller and therefore cheaper and more quickly too.

But the key point is not that any one company has the single breakthrough idea we need in nuclear fission or fusion. What's most important is that the world get serious once again about advancing the field of nuclear energy. It's just too promising to ignore.

Offshore wind. Putting wind turbines in an ocean or other body of water has various advantages. Because many major cities are near the coast, we can generate electricity much closer to the places where it'll be used and not run into as many transmission problems. Offshore winds generally blow more steadily, so intermittency is less of an issue too.

Despite these advantages, offshore wind currently represents only a tiny share of the world's total capacity for generating electricity—about 0.4 percent in 2019. Most of that is in Europe, particularly in the North Sea; the United States has just 30 megawatts installed, and that's all in one project off the coast of Rhode Island. Remember that America uses around 1,000 gigawatts, so offshore wind provides roughly 1/32,000th of the country's electricity.

For the offshore wind industry, there's nowhere to go but up. Companies are finding ways to make turbines bigger so each one can generate more power, and they're solving some of the engineering challenges involved in placing large metal objects out in the ocean. As these innovations drive down the price, countries are installing more turbines; the use of offshore wind has grown at an average annual rate of 25 percent in the past three years. The U.K. is the world's biggest user of offshore wind today, thanks to clever government subsidies that encouraged companies to invest in it. China is making big investments in offshore wind and will likely be the world's biggest consumer of it by 2030.

The United States has considerable offshore wind available, especially in New England, Northern California and Oregon, the Gulf Coast, and the Great Lakes; in theory, we could generate 2,000 gigawatts from it—more than enough to meet our current needs. But if we're going to take advantage of this potential, we'll have to make it easier to put up turbines. Today, getting a permit requires you to run a bureaucratic gauntlet: You buy one of a limited number of federal leases, then go through a multiyear process to generate an environmental impact statement, then get additional state and local permits. And at each step of the way, you may be opposed (rightly or not) by beachfront property owners, the tourism industry, fishermen, and environmental groups.

Offshore wind holds a lot of promise: It's getting cheaper and can play a key role in helping many countries decarbonize.

Geothermal. Deep underground—as close as a few hundred feet, as far down as a mile—are hot rocks that can be used to generate carbon-free electricity. We can pump water at high pressure down into the rocks, where it absorbs the heat and then comes out another hole, where it turns a turbine or generates electricity some other way.

But exploiting the heat under our feet has its downsides. Its energy density—the amount of energy we get per square meter—is quite low. In his fantastic 2009 book, *Sustainable Energy—Without the Hot Air,* David MacKay estimated that geothermal could meet less than 2 percent of the U.K.'s energy needs, and delivering even that much would require exploiting every square meter of the country and doing the drilling for free.

We also have to dig wells to reach it, and it's hard to know ahead of time whether any given well is going to produce the heat we need, or for how long. Some 40 percent of all wells dug for geothermal turn out to be duds. And geothermal is available only in certain places around the world; the best spots tend to be areas with above-average volcanic activity.

Although these problems mean that geothermal will contribute only modestly to the world's power consumption, it's still worth setting out to solve them one by one, just as we did with cars. Companies are working on various innovations that would build on the technical advances that have made oil and gas drilling so much more productive in the past few years. For example, some are developing advanced sensors that could make it easier to find promising geothermal wells. Others are using horizontal drills so they can tap these geothermal sources more safely and efficiently. It's a great example of how technology that was originally developed for the fossil-fuel industry can actually help drive us toward zero emissions.

Storing Electricity

Batteries. I've spent way more time learning about batteries than I ever would've imagined. (I've also lost more money on start-up battery companies than I ever imagined.) To my surprise, despite all the limitations of lithium-ion batteries—the ones that power your laptop and mobile phone—it's hard to improve on them. Inventors have studied all the metals we could use in batteries, and it seems unlikely that there are materials that will make for vastly better batteries than the ones we're already building. I think we can improve them by a factor of 3, but not by a factor of 50.

Still, you can't keep a good inventor down. I've met some brilliant engineers working on affordable batteries that could store enough energy for a city—what we call grid-scale batteries, as opposed to the smaller ones that run a phone or computer—and hold it long enough to get through seasonal intermittency. One inventor I admire is working on a battery that uses liquid metals instead of the solid metals employed in traditional batteries. The idea is that liquid metal lets you store and deliver much more energy very quickly— exactly the kind of thing you need when you're trying to power an entire city. The technology has been proven in a lab, and now the team is trying to make it cheap enough to be economical and prove that it works in the field.

Others are working on something called flow batteries, which involve storing fluids in separate tanks and then generating electricity by pumping the fluids together. The bigger the tanks, the more energy you can store, and the bigger the battery, the more economical it becomes.

Pumped hydro. This is a method of storing city-sized amounts of energy, and it works like this: When electricity is cheap (for example, when a stiff wind is turning your turbines really fast), you pump water up a hill into a reservoir; then, when demand for power goes up, you let the water flow back down the hill, using it to spin a turbine and generate more electricity.

Pumped hydro is the biggest form of grid-scale electricity storage in the world. Unfortunately, that's not saying much. The 10 largest facilities in the United States can store less than an hour's worth of the country's electricity consumption. You can probably guess why it hasn't really taken off: To pump water up a hill, you need a big reservoir of water and, of course, a hill. Without either, you're out of luck.

Several companies are working on alternatives. One is looking at whether you could move something other than water uphill—pebbles, for example. Another is working on a process that would do away with the hill but not the water: You pump water underground, keep it there under pressure, and then release it when you're ready to turn a turbine. If this approach works, it would be magical, because there would be very little aboveground equipment to worry about.

Thermal storage. The notion here is that when electricity is cheap, you use it to heat up some material. Then, when you need more electricity, you use the heat to generate power via a heat engine. This can work at 50 or 60 percent efficiency, which isn't bad. Engineers know about many materials that can stay hot for a long time without losing much energy; the most promising approach, which some scientists and companies are working on, is to store the heat in molten salt.

At TerraPower, we're trying to figure out how to use molten salt so that (if we're able to build a plant) we don't have to compete with solar-generated electricity during the day. The idea would be to store heat generated during the day, then convert it to electricity at night, when cheap solar power isn't available.

Cheap hydrogen. I hope we get some big breakthroughs in storage. But it's also possible that some innovation will come along and make all these ideas obsolete, the way the personal computer came along and more or less made the typewriter unnecessary.

Cheap hydrogen could do that for storing electricity.

The reason is that hydrogen serves as a key ingredient in fuel cell batteries. Fuel cells get their energy from a chemical reaction between two gases—usually hydrogen and oxygen—and their only by-product is water. We could use electricity from a solar or wind farm to create hydrogen, store the hydrogen as compressed gas or in another form, and then put it in a fuel cell to generate electricity on demand. In effect, we'd be using clean electricity to create a carbon-free fuel that could be stored for years and turned back into electricity at a moment's notice. And we would solve the location problem I mentioned earlier; although you can't ship sunlight in a railcar, you can turn it into fuel first and then ship it any way you like.

Here's the problem: Right now, it's expensive to produce hydrogen without emitting carbon. It's not as efficient as storing the electricity directly in a battery, because first you have to use electricity to make hydrogen and then later you use that hydrogen to make electricity. Taking all these steps means you lose energy along the way.

Hydrogen is also a very lightweight gas, which makes it hard to store within a reasonably sized container. It's easier to store the gas if you pressurize it (you can squeeze more into the same-volume container), but because hydrogen molecules are so small, when they're under pressure, they can actually migrate through metals. It's as if your gas tank slowly leaked gas as you filled up.

Finally, the process of making hydrogen (called electrolysis) also requires various materials (known as electrolyzers) that are quite costly. In California, where cars that run on fuel cells are now available, the cost of hydrogen is equivalent to paying $5.60 a gallon for gasoline. So scientists are experimenting with cheaper materials that could serve as electrolyzers.

Other Innovations

Capturing carbon. We could keep making electricity as we do now, with natural gas and coal, but suck up the carbon dioxide before it hits the atmosphere. That's called carbon capture and storage, and it involves installing special devices at fossil-fuel plants to absorb emissions. These "point capture" devices have existed for decades, but they're expensive to buy and operate, they generally capture only 90 percent of the greenhouse gases involved, and power companies don't gain anything from installing them. So very few are in use. Smart public policies could create incentives to use carbon capture, a subject we'll return to in chapters 10 and 11.

Earlier, I mentioned a related technology called direct air capture. It involves exactly what the name implies: capturing carbon directly from the air. DAC is more flexible than point capture, because you can do it anywhere. And in all likelihood, it'll be a crucial part of getting to zero; one study by the National Academy of Sciences found that we'll need to be removing about 10 billion tons of carbon dioxide a year by mid-century and about 20 billion by the end of the century.

But DAC is a much bigger technical challenge than point capture, thanks to the low concentration of carbon dioxide in the air. When emissions come directly out of a coal plant, they're highly concentrated, in the range of 10 percent carbon dioxide, but once they're in the atmosphere, where DAC operates, they disperse widely. Pick one molecule at random out of the atmosphere and the odds that it will be carbon dioxide are just 1 in 2,500.

Companies are working on new materials that are better at absorbing carbon dioxide, which will make both point capture and DAC cheaper and more effective. In addition, today's approaches to DAC require a lot of energy to trap the greenhouse gases, collect them, and store them safely. There's no way to do all that work without using *some* energy; the laws of physics set a minimum amount on how much will be required. But the latest technology uses much more than that minimum, so there's a lot of room for improvement.

Using less. I used to scoff at the notion that using power more efficiently would make a dent in climate change. My rationale: If you have limited resources to reduce emissions (and we do), then you'd get the biggest impact by moving to zero emissions rather than by spending a lot trying to reduce the demand for energy.

I haven't abandoned that view entirely, but I did soften it when I realized just how much land it will take to generate lots more electricity from solar and wind. A solar farm needs between 5 and 50 times more land to generate as much electricity as an equivalent coal-powered plant, and a wind farm needs 10 times more than solar. We should do everything we can to increase the odds that we can scale up to 100 percent clean power, and that will be easier if we reduce electricity demand wherever we can. Anything that reduces the scale we need to reach is helpful.

There's also a related approach called load shifting or demand shifting, which involves using power more consistently throughout the day. If we did it on a large scale, load shifting would represent a pretty big change in the way we think about powering our lives. Right now, we tend to generate power when we use it—for example, cranking up electric plants to run a city's lights at night. With load shifting, though, we do the opposite: We use more electricity when it's cheapest to generate.

For example, your water heater might be able to switch on at 4:00 p.m., when power is less in demand, instead of 7:00 p.m. Or you could plug in your electric vehicle when you get home for the day, and it would automatically wait to charge itself until 4:00 a.m., when electricity is cheap because so few people are using it. On an industrial level, energy-intensive processes like treating wastewater and making hydrogen fuels could be done at a time of day when power is easiest to come by.

If load shifting is going to have a significant impact, we'll need some changes in policy as well as some technological advances. Utility companies will have to update the price of electricity throughout the day to account for shifts in supply and demand, for instance, and your water heater and electric car will have to be smart enough to take advantage of this price information and respond accordingly. And in extreme cases, when electricity is especially hard to come by, we should have the ability to shed demand, meaning we'd ration electricity, prioritize the highest needs (say, hospitals), and shut down nonessential activities.

Keep in mind that although we need to pursue all these ideas, we probably don't need all of them to pan out in order to decarbonize our power grid. Some of the ideas overlap each other. If we get a breakthrough in cheap hydrogen, for example, we might not need to worry as much about getting a magic battery.

What I can say for certain is that we need a concrete plan to develop new power grids that provide affordable zero-carbon electricity reliably, whenever we need it. If a genie offered me one wish, a single breakthrough in just one activity that drives climate change, I'd pick making electricity: It's going to play a big role in decarbonizing other parts of the physical economy. I'll turn to the first of these—how we make things like steel and cement—in the next chapter.

CHAPTER 5

HOW WE MAKE THINGS

29 percent of 52 billion tons per year

I t's an eight-mile drive from Medina, Washington, where Melinda and I live, to the Seattle headquarters of our foundation. To get to the office, I cross Lake Washington on what's officially known as the Evergreen Point Floating Bridge, although no one who lives around here actually calls it that; to locals, it's the 520 bridge, named for the state highway that runs across it. At more than 7,700 feet, it's the longest floating bridge in the world.

Every so often when I cross the 520 bridge, I take a moment to appreciate how marvelous it is. Not because it's the longest floating bridge in the world, but because *it's a bridge that floats*. How can this massive structure made with tons of asphalt, concrete, and steel, and with hundreds of cars sitting on it, float on top of a lake? Why the hell doesn't it sink?

The answer is a miracle of engineering brought to us by an amazing material: concrete. At first glance, this may seem strange, because it's so natural to think of concrete as this heavy block that couldn't possibly float. Although it's true that concrete can be made that way—solid enough to absorb nuclear radiation in the walls of a hospital—it can also be used to make hollow shapes, like the 77 air-filled, watertight pontoons that support the 520 bridge. Each weighs thousands of tons, is buoyant enough to float on the surface

This is the 520 bridge in Seattle, which I cross whenever I drive from home to the Gates Foundation's headquarters. It's a marvel of modern engineering.

of the lake, and is sturdy enough to support the bridge and all the cars speeding across it. Or, more likely, inching across it, during one of our daily traffic jams.

You don't have to look very hard to find concrete performing other miracles around you. It's rust-resistant, rot-proof, and non-flammable, which is why it's part of most modern buildings. If you're a fan of hydropower, you should appreciate concrete for making dams possible. The next time you see the Statue of Liberty, take a look at the pedestal she's standing on. It's made of 27,000 tons of concrete.

The charms of concrete were not lost on America's greatest inventor. Thomas Edison tried to create entire homes built out of the stuff. He dreamed of making concrete furniture, like bedroom sets, and even tried to design a concrete record player.

These imaginings of Edison's never came to pass, but even so we use a *lot* of concrete. Every year, between replacing or repairing existing roads, bridges, and buildings and putting up new ones, America alone produces more than 96 million tons of cement, one of the main ingredients in concrete. That's nearly 600 pounds for every person in the country. And we're not even the biggest consumers of the stuff—that would be China, which installed more concrete in the first 16 years of the 21st century than the United States did in the entire 20th century!

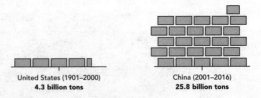

United States (1901–2000)
4.3 billion tons

China (2001–2016)
25.8 billion tons

China makes a lot of cement. The country has already produced more in the 21st century than the United States did in the entire 20th century. (U.S. Geological Survey)

Obviously, cement and concrete aren't the only materials we rely on. There's also the steel we put in cars, ships, and trains; refrigerators and stoves; factory machines; cans of food; and even computers. Steel is strong, cheap, durable, and infinitely recyclable. It also makes a terrific partner with concrete: Insert steel rods inside a block of concrete, and you've got a magical construction material that can withstand tons of weight and also won't break apart when you twist it. That's why we use reinforced concrete in most of our buildings and bridges.

Americans use as much steel as cement—so that's another 600 pounds per person, every year, not counting the steel we recycle and use again.

Plastics are another amazing material. They're in so many products, from clothes and toys to furniture and cars and cell phones, that it's impossible to list them all. Plastics have a bad reputation these days, a reputation that's partially fair. But they also do a lot of good. As I write this chapter, I'm sitting at my desk and can see plastics all around me: my computer, keyboard, monitor, and mouse, my stapler, my phone, and on and on. Plastics are also what allow fuel-efficient cars to be so light; they account for as much as half of a car's total volume, but only 10 percent of its weight.

Then there's glass—in our windows, jars and bottles, insulation, cars, and the fiber-optic cables that give you a high-speed internet connection. Aluminum goes into soda cans, foil, power lines, doorknobs, trains, planes, and beer kegs. Fertilizer helps feed the world. Years ago, I predicted the demise of paper as electronic communications became more common and screens became more ubiquitous, but it doesn't show much sign of going away anytime soon.

In short, we make materials that have become just as essential to modern life as electricity is. We're not going to give them up. If anything, we'll be using more of them as the world's population grows and gets richer.

There's copious data to back up this claim—we'll be producing 50 percent more steel by mid-century than we do today, for example—but I think the two pictures below are just as persuasive.

Take a quick look at them. They look like two different cities, right?

They aren't. They're both photos of Shanghai, taken from the

These two photos capture what growth looks like—for better and for worse. Shanghai in 1987 (left) and 2013 (right).

same vantage point. The one on the left was taken in 1987, the one on the right in 2013. When I look at all those new buildings in the photo on the right, I see tons and tons of steel, cement, glass, and plastic.

This story is being repeated all over the world, though the growth in most places isn't as dramatic as it was in Shanghai. To repeat a theme that comes up repeatedly in this book: *This progress is a good thing.* The rapid growth you see in these two photos means that people's lives are improving in countless ways. They are earning more money, are getting a better education, and are less likely to die young. Anyone who cares about fighting poverty should see it as good news.

But, to repeat another theme that comes up a lot in this book: *This silver cloud has a dark lining.* Making all these materials emits lots of greenhouse gases. In fact, they're responsible for about a third of all emissions worldwide. And in some cases, notably concrete, we don't have a practical way to make them without producing carbon.

So let's look at how we can square this circle—how we can keep producing these materials without making the climate unlivable. For the sake of brevity, we'll focus on three of the most important materials: steel, concrete, and plastic. As we did with electricity, we'll look at how we got here and why these materials are so problematic for the climate. Then we'll calculate the Green Premiums for reducing emissions using today's technology, and we'll examine ways to drive down the Green Premiums and make all this stuff without emitting carbon.

The history of steel goes back some 4,000 years. There's a long series of fascinating inventions over the centuries that got us from the Iron Age to the cheap, versatile steel we have today, but in my experience most people don't want to hear a lot about the differences between blast furnaces, puddling furnaces, and the Bessemer process. So here are the main things you need to know.

We like steel because it's both strong and easy to shape when it's hot. To make steel, you need pure iron and carbon; on its own, iron isn't very strong, but add just the right amount of carbon—less than 1 percent, depending on the kind of steel you want—and the carbon atoms nestle themselves in between the iron atoms, giving the resulting steel its most important properties.

Carbon and iron aren't hard to find—you can get carbon from coal, and iron is a common element in the earth's crust. But pure iron is quite rare: When you dig up the metal, it's almost always combined with oxygen and other elements—a mixture known as iron ore.

To make steel, you need to separate the oxygen from the iron and add a tiny bit of carbon. You can accomplish both at the same time by melting iron ore at very high temperatures (1,700 degrees Celsius or over 3,000 degrees Fahrenheit), in the presence of oxygen and a type of coal called coke. At those temperatures, the iron ore releases its oxygen, and the coke releases its carbon. A bit of the carbon bonds with the iron, forming the steel we want, and the rest of the carbon grabs onto the oxygen, forming a by-product we don't want: carbon dioxide. Quite a bit of carbon dioxide, in fact. Making 1 ton of steel produces about 1.8 tons of carbon dioxide.

Why do we do it this way? Because it's cheap, and until we started worrying about climate change, we had no incentive to do it any other way. Iron ore is pretty easy (and therefore inexpensive) to dig up. Coal too is inexpensive, because there's so much of it in the ground.

So the world is going to keep chugging along, making more steel, even as production basically plateaus in the United States. Several other countries now produce more raw steel than the United States does—China, India, and Japan among them—and by 2050 the world will be producing roughly 2.8 billion tons every year. That adds up to 5 billion tons of carbon dioxide released every year by mid-century, just from making steel, unless we find a new, climate-friendly way to do it.

As challenging as that may sound, concrete is even harder. (Sorry—no pun intended.) To make it, you mix together gravel, sand, water, and cement. The first three of these are relatively easy; it's the cement that is a problem for the climate.

To make cement, you need calcium. To get calcium, you start with limestone—which contains calcium plus carbon and oxygen—and burn it in a furnace along with some other materials.

Given the presence of carbon and oxygen, you can probably see where this is going. After burning the limestone, you end up with the thing you want—calcium for your cement—plus something you don't want: carbon dioxide. Nobody knows of a way to make cement without going through this process. It's a chemical reaction—*limestone plus heat* equals *calcium oxide plus carbon dioxide*—and there's no way around it. It's a one-to-one relationship. Make a ton of cement, and you'll get a ton of carbon dioxide.

And, just like with steel, there's no reason to think we're going to stop making cement. China is the biggest producer by far, outpacing second-place India by a factor of seven and making more than the rest of the world combined. Between now and 2050, the world's annual cement production will go up a bit—as the building boom slows in China and picks up in smaller developing countries—before settling back down near 4 billion tons a year, roughly where it is today.

Compared with cement and steel, plastics are the baby of the group. Although humans were using natural plastics, such as rubber, thousands of years ago, synthetic plastics only came into their own in the 1950s, thanks to some breakthroughs in chemical engineering. Today there are more than two dozen types of plastics, and they range from the kind of thing you might expect—the polypropyl-ene in yogurt containers, for example—to more surprising uses like the acrylic in paint, floor polish, and laundry detergent, or the microplastics in soap and shampoo, or the nylon in your waterproof jacket, or the polyester in all those regrettable clothes I wore in the 1970s.

All these different types of plastics have one thing in common: They contain carbon. Carbon, it turns out, is useful in creating all sorts of different materials because it bonds easily with a wide variety of different elements; in the case of plastics, it's usually clustered with hydrogen and oxygen.

Since you've read this far, you probably won't be surprised to learn where companies that make plastics tend to get their carbon. They get it by refining oil, coal, or natural gas and then processing the refined products in various ways. This helps explain why plastics have earned a reputation for being inexpensive: Like cement and steel, plastics are cheap because fossil fuels are cheap.

But there's one important way that plastics are fundamentally different from cement and steel. When we make cement or steel, we release carbon dioxide as an inevitable by-product, but when we make a plastic, around half of the carbon stays in the plastic. (The actual percentage varies quite a bit, depending on which kind of plastic you're talking about, but around half is a reasonable approxi-mation.) Carbon really likes bonding with the oxygen and hydrogen, and it isn't inclined to let go. Plastics can take hundreds of years to degrade.

That's a major environmental problem, because the plastics that get dumped in landfills and oceans stick around for a century or more. And it's a problem that's worth solving: Pieces of plastic floating around in the ocean cause all sorts of problems, including poisoning marine life. But they're not making climate change worse. Purely in terms of emissions, the carbon in plastics is not such bad news. Because plastics take so long to degrade, all the carbon atoms that go into them are atoms that won't go into the atmosphere and drive up the temperature—at least not for a very long time.

I'll pause here to emphasize that this quick survey covers only three of the most important materials we make today. I'm leaving out fertilizer, glass, paper, aluminum, and many others. But the key points remain the same: We manufacture an enormous amount of materials, resulting in copious amounts of greenhouse gases, nearly a third of the 52 billion tons per year. We need to get those emissions down to zero, but it's not an option to simply stop making things. In the rest of this chapter, we'll examine the alternatives, see how high the Green Premiums are, and then look at how technology might drive the premiums down so everyone will want to adopt the zero-emissions approach.

To figure the Green Premiums on materials, you need to understand where emissions come from when we make things. I think of it in three stages: We emit greenhouse gases (1) when we use fossil fuels to generate the electricity that factories need to run their operations; (2) when we use them to generate heat needed for different manufacturing processes, like melting iron ore to make steel; and (3) when we actually make these materials, like the way cement manufacturing inevitably creates carbon dioxide. Let's take these one by one and see how they contribute to the Green Premiums.

For the first stage, electricity, we covered most of the key challenges in chapter 4. After you factor in storage and transmission, and the fact that many factories need reliable power around the clock, the cost for clean electricity goes up fast—much more for most countries than for the United States or Europe.

Then there's the second stage: How can we generate heat without burning fossil fuels? If you don't need super-high temperatures, you can use electric heat pumps and other technologies. But when you're looking for temperatures in the thousands of degrees, electricity isn't an economical option—at least not with today's technology. You'll have to either use nuclear power or burn fossil fuels and grab the emissions with carbon-capture devices. Unfortunately, carbon capture doesn't come for free. It adds to the manufacturer's cost and gets passed on to the consumer.

Finally, the third stage: What can we do about the processes that inherently produce greenhouse gas emissions? Remember that making steel and cement emits carbon dioxide—not just from burning fossil fuels, but as a result of the chemical reactions that are essential to their creation.

Right now, the answer is clear: Short of simply shutting down these parts of the manufacturing sector, we can do nothing today to avoid these emissions. If we wanted to go all in on eliminating them using whatever technologies we have available today, our options would be as limited as they were in the second stage. We'd have to use fossil fuels and carbon capture—which, again, adds to the cost.

With those three stages in mind, let's look at the range of Green Premiums for using carbon capture to make clean plastics, steel, and cement:

Green Premiums for plastics, steel, and cement

Material	Average price per ton	Carbon emitted per ton of material made	New price after carbon capture	Green Premium range
Ethylene (plastic)	$1,000	1.3 tons	$1,087–$1,155	9%–15%
Steel	$750	1.8 tons	$871–$964	16%–29%
Cement	$125	1 ton	$219–$300	75%–140%

Aside from cement, these premiums may not seem like much. And it's true that in some cases, consumers might not feel any pinch at all. For example, a $30,000 car might contain one ton of steel; whether this steel costs $750 or $950 hardly makes any difference in the overall price of the car. Even for that $2 bottled Coke you bought out of a vending machine the other day, the plastic represents a minuscule share of the overall price.

But the final cost to consumers isn't the only factor that matters. Suppose you're an engineer working for the City of Seattle, and you're reviewing bids to repair one of our many bridges. One bid comes in charging $125 a ton for cement, and another comes in charging $250 a ton, having added on the cost for carbon capture. Which one will you pick? Without some incentive to opt for the zero-carbon cement, you'll go with the cheaper one.

Or, if you run a car company, will you be willing to spend 25 percent more on all the steel you buy? Probably not, especially if your competitors decide to stick with the cheaper stuff. The fact that the overall price of the car will increase only a tiny bit wouldn't be much comfort to you. Your margins are already pretty slim, and you'd be unhappy to see the price of one of your most important commodities go up by a quarter. In an industry with narrow profit margins, a 25 percent premium could be the difference between staying in business and going broke.

Although a few manufacturers in a few industries might be willing to bear the cost for the right to say they're doing their part to fight climate change, at these prices we'll never drive the kind of system-wide change we need to get to zero. Nor can we count on consumers to drive down the prices by demanding more of these green products. After all, consumers don't buy cement or steel—large corporations do.

There are different ways to bring the premiums down. One is by using public policies to create demand for clean products—for example, by creating incentives or even requirements to buy zero-carbon cement or steel. Businesses are much more likely to pay the premium for clean materials if the law requires it, their customers demand it, and their competitors are doing it. I'll cover these incentives in chapters 10 and 11.

But—and this is essential—we need innovation in the manufacturing process, ways to make things without emitting carbon. Let's look at some of the opportunities.

Of all the materials I've covered in this chapter, cement is the toughest case of all. It's hard to get around that simple fact—*limestone plus heat* equals *calcium oxide plus carbon dioxide*. But a number of companies have good ideas.

One approach is to take recycled carbon dioxide—possibly captured during the process of making cement—and inject it back into the cement before it's used at the construction site. The company that's pursuing this idea has several dozen customers already, including Microsoft and McDonald's; so far it's only able to reduce emissions by around 10 percent, though it hopes to get to 33 percent eventually. Another, more theoretical approach involves making cement out of seawater and the carbon dioxide captured from power plants. The inventors behind this idea think it could ultimately cut emissions by more than 70 percent.

Yet even if these approaches are successful, they won't give us 100 percent carbon-free cement. For the foreseeable future, we'll have to count on carbon capture and—if it becomes practical—direct air capture to grab the carbon emitted when we make cement.

For pretty much all other materials, the first thing we need is plenty of *reliable clean electricity*. Electricity already accounts for about a quarter of all the energy used by the manufacturing sector worldwide; to power all these industrial processes, we need to both deploy the clean energy technology we already have and develop breakthroughs that let us generate and store lots of zero-carbon electricity inexpensively.

And soon we'll need even more power, as we pursue another way to reduce emissions: *electrification,* which is the technique of using electricity instead of fossil fuels for some industrial processes. For example, one very cool approach for steelmaking is to use clean electricity to replace coal. A company I'm following closely has developed a new process called molten oxide electrolysis: Instead of burning iron in a furnace with coke, you pass electricity through a cell that contains a mixture of liquid iron oxide and other ingredients. The electricity causes the iron oxide to break apart, leaving you with the pure iron you need for steel, and pure oxygen as a by-product. No carbon dioxide is produced at all. This technique is promising—it's similar to a process we've been using for more than a century to purify aluminum—but like the other ideas for clean steel it hasn't yet been proven to work at an industrial scale.

Clean electricity would help us solve another problem too: making plastics. If enough pieces come together, plastics could one day become a carbon sink—a way to remove carbon rather than emit it.

Here's how we could do it. First, we would need a zero-carbon way to power the refining process. We could do that with clean electricity or with hydrogen produced from clean electricity. Then we'd need a way to get the carbon for our plastics without burning coal. One idea is to remove carbon dioxide from the air and extract the carbon, though that's an expensive process. Another approach that various companies are working on is to get carbon from plants. Finally, we'd need a zero-carbon source of heat—which would likely also be clean electricity, hydrogen, or natural gas fitted with a device to capture the carbon it emits.

If all these pieces come together, we could make plastics with net-negative emissions. In effect, we'd find a way to take carbon out of the air (using plants or some other method) and put it into a bottle or other plastic product, where it would stay for decades or centuries, with no additional emissions. We'd be socking away more carbon than we were putting out.

Beyond finding ways to make materials with zero emissions, we can simply use less stuff. On its own, recycling more of our steel, cement, and plastic won't be nearly enough to eliminate greenhouse gas emissions, but it will help. We can recycle more materials and should be exploring new ways to cut the amount of energy needed to recycle stuff. And because reusing something doesn't require nearly as much energy as recycling it, we should also be looking at ways to build and make things using repurposed materials. Finally, buildings and roads can also be designed with the goal of limiting the use of cement and steel, and in some cases cross-laminated wood—made of layers of timber glued together into a stack—is sturdy enough to substitute for both materials.

To sum up, the path to zero emissions in manufacturing looks like this:

1. *Electrify every process possible.* This is going to take a lot of innovation.
2. *Get that electricity from a power grid that's been decarbonized.* This also will take a lot of innovation.
3. *Use carbon capture to absorb the remaining emissions.* And so will this.
4. *Use materials more efficiently.* Same.

Get used to this theme. You'll see it often in the coming chapters. Next up is agriculture, which features one of the great unsung heroes of the 20th century as well as farms full of burping cows.

CHAPTER 6

HOW WE GROW THINGS

22 percent of 52 billion tons a year

Cheeseburgers run in my family. When I was a kid, I'd go on hikes with my Boy Scout troop, and all the guys always wanted to ride home with my dad because he'd stop along the way and treat everyone to burgers. Many years later, in the early days of Microsoft, I scarfed down countless lunches, dinners, and late-night meals at the nearby Burgermaster, one of the Seattle area's oldest burger chains.

Eventually, after Microsoft became successful but before Melinda and I started our foundation, my dad started using the Burgermaster near his house as an unofficial office. He'd sit in the restaurant, eating lunch while he sifted through requests we had received from people who were asking for donations. After a while, word got out, and Dad started getting letters addressed to him there: "Bill Gates Sr., in care of Burgermaster."

Those days are long gone. It's been two decades since Dad traded in his table at Burgermaster for a desk at our foundation. And although I still love a good cheeseburger, I don't eat them nearly as often as I used to—because of what I've learned about the impact that beef and other meats have on climate change.

Raising animals for food is a major contributor of greenhouse gas emissions; it ranks as the highest contributor in the sector that experts call "agriculture, forestry, and other land use," which in turn covers a huge range of human activity, from raising animals and growing crops to harvesting trees. This sector also involves a wide range of various greenhouse gases: With agriculture, the main culprit isn't carbon dioxide but methane—which causes 28 times more warming per molecule than carbon dioxide over the course of a century—and nitrous oxide, which causes *265 times* more warming.

All told, each year's emissions of methane and nitrous oxide are the equivalent of more than 9 billion tons of carbon dioxide, or more than 80 percent of all the greenhouse gases in this ag/forestry/land use sector. Unless we do something to curb these emissions, that number will go up as we grow enough food to feed a global population that's getting bigger and richer. If we want to get near net-zero emissions, we're going to have to figure out how to grow plants and raise animals while reducing and eventually eliminating greenhouse gases.

And farming isn't the only challenge. We'll also have to do something about deforestation and other uses of land, which together add hundreds of millions of tons of carbon dioxide to the atmosphere while also destroying essential wildlife habitats.

In keeping with such a wide-ranging subject, this chapter has a bit of everything. I'll tell you about one of my heroes, a Nobel Peace Prize–winning agronomist who saved a billion people from starvation but whose name is largely unknown outside global-development circles. We'll also explore the ins and outs of pig manure and cow burps, the chemistry of ammonia, and whether planting trees helps avoid a climate disaster. But before we get to any of that, let's start with a famous prediction that turned out to be historically wrong.

In 1968, an American biologist named Paul Ehrlich published a best-selling book called *The Population Bomb,* in which he painted a grim picture of the future that was not far removed from the dystopian vision of novels like *The Hunger Games*. "The battle to feed all of humanity is over," Ehrlich wrote. "In the 1970s and 1980s hundreds of millions of people will starve to death in spite of any crash programs embarked upon now." Ehrlich also wrote that "India couldn't possibly feed 200 million more people by 1980."

None of this came to pass. In the time since *The Population Bomb* came out, India's population has grown by more than 800 million people—it's now more than double what it was in 1968—but India produces more than three times as much wheat and rice as it did back then, and its economy has grown by a factor of 50. Farmers in many other countries throughout Asia and South America have seen similar productivity gains.

As a result, even though the global population is growing, there are not hundreds of millions of people starving to death in India or anywhere else. In fact, food is becoming more affordable, not less. In the United States, the average household spends less of its budget today on food than it did 30 years ago, a trend that's being repeated in other parts of the world as well.

I'm not saying that malnutrition isn't a serious problem in some places. It is. In fact, improving nutrition for the world's poorest is a key priority for Melinda and me. But Ehrlich's prediction of mass starvation was wrong.

Why? What did Ehrlich and other doomsayers miss?

They didn't factor in the power of innovation. They didn't account for people like Norman Borlaug, the brilliant plant scientist who sparked a revolution in agriculture that led to the gains in India and elsewhere. Borlaug did it by developing varieties of wheat with bigger grains and other characteristics that allowed them to provide much more food per acre of land—what farmers call raising the yield. (Borlaug found that as he made the grains bigger, the wheat couldn't stand up under their weight, so he made the wheat stalks shorter, which is why his varieties are known as semi-dwarf wheat.)

As Borlaug's semi-dwarf wheat spread around the world, and as other breeders did similar work on corn and rice, yields tripled in most areas. Starvation plummeted, and today Borlaug is widely credited with saving a billion lives. He won the Nobel Peace Prize in 1970, and we're still feeling the impact of his work: Virtually all the wheat grown on earth is descended from the plants he bred. (One downside of these new varieties is that they need lots of fertilizer to reach their full growth potential, and as we'll discuss in a later section, fertilizer has some negative side effects.) I love the fact that one of history's greatest heroes had a job title—agronomist—that most of us have never even heard of.

So what does Norman Borlaug have to do with climate change?

The global population is headed toward 10 billion people by 2100, and we're going to need more food to feed everyone. Because we'll have 40 percent more people by the end of the century, it would be natural to think that we'll need 40 percent more food too, but that's not the case. We'll need even more than that.

Here's why: As people get richer, they eat more calories, and in particular they eat more meat and dairy. And producing meat and dairy will require us to grow even more food. A chicken, for example, has to eat two calories' worth of grain to give us one calorie of poultry—that is, you have to feed a chicken twice as many calories as you'll get from the chicken when you eat it. A pig eats three times as many calories as we get when we eat it. For cows, the ratio is highest of all: six calories of feed for every calorie of beef. In other words, the more calories we get from these meat sources, the more plants we need to grow for the meat.

This chart shows you the trends in meat consumption around the world. It's basically flat in the United States, Europe, Brazil, and Mexico, but it's climbing rapidly in China and other developing countries.

Here's the conundrum: We need to produce much more food

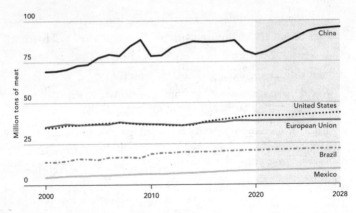

Most countries aren't consuming more meat than they used to. China, though, is a big exception. (OECD-FAO Agricultural Outlook 2020)

than we do today, but if we keep producing it with the same methods we use now, it will be a disaster for the climate. Assuming we don't make any improvements in the amount of food we get per acre of pasture or cropland, growing enough to feed 10 billion people will drive up food-related emissions by two-thirds.

Another concern: If we make a big push to generate energy from plants, we could accidentally spark a competition for cropland. As I'll describe in chapter 7, advanced biofuels made from things like switchgrass could give us zero-carbon ways to power trucks, ships, and airplanes. But if we grow those crops on land that would otherwise be used to feed a growing population, we could inadvertently drive up food prices, pushing even more people into poverty and malnutrition while accelerating the already dangerous pace of deforestation.

To avoid these traps, we're going to need more Borlaug-sized breakthroughs in the years ahead. Before we can look at what those breakthroughs might be, though, I want to explain where exactly all these emissions are coming from and explore our options for eliminating them using today's technology. Just as I did in the previous chapter, I'll use Green Premiums to show why getting rid of these greenhouse gases is too expensive today, and to make the case that we need some new inventions.

Which brings me to cow burps and pig manure.

Look inside a person's stomach and you'll find just one chamber where food starts getting digested before making its way to the intestinal tract. But look inside a cow's stomach, and you'll see four chambers. These compartments are what allow the cow to eat grass and other plants that humans can't digest. In a process called enteric fermentation, bacteria inside the cow's stomach break down the cellulose in the plant, fermenting it and producing methane as a result. The cow belches away most of the methane, though a little comes out the other end as flatulence.

(By the way, when you get into this subject, you can end up having some weird conversations. Each year, Melinda and I publish an open letter about our work, and in our 2019 letter I decided to write about this problem of enteric fermentation in cattle. One day, as we were going over a draft, Melinda and I had a healthy debate about how many times I could use the word "fart" in the letter. She got me down to one. As the sole author of this book, I have more leeway, and I intend to use it.)

Around the world, there are roughly a billion cattle raised for beef and dairy. The methane they burp and fart out every year has the same warming effect as 2 billion tons of carbon dioxide, accounting for about 4 percent of all global emissions.

Burping and farting natural gas is a problem that's unique to cows and other ruminants, like sheep, goats, deer, and camels. But there's another cause of greenhouse gas emissions that's common to every animal: poop.

When poop decomposes, it releases a mix of powerful greenhouse gases—mostly nitrous oxide, plus some methane, sulfur, and ammonia. About half of poop-related emissions come from pig manure, and the rest from cow manure. There's so much animal poop that it's actually the second-biggest cause of emissions in agriculture, behind enteric fermentation.

What can we do about all this pooping, burping, and farting? That's a tough one. Researchers have tried all sorts of ideas for dealing with enteric fermentation. They've tried using vaccines to cut down on the methanogenic microbes living in the cattle's gut, breeding cattle to naturally produce fewer emissions, and adding special feeds or drugs to their diets. These efforts have mostly been unsuccessful, though one promising exception is a compound called 3-nitrooxypropanol, which reduces methane emissions by 30 percent. But right now you have to give it to the cattle at least once a day, so it's not yet feasible for most grazing operations.

Still, there's reason to believe we can cut down on these emissions without any new technology and without a significant Green Premium. It turns out the amount of methane produced by a given cow depends a lot on where the cow lives; for example, cattle in South America emit up to five times more greenhouse gases than ones in North America do, and African cattle emit even more. If a cow is being raised in North America or Europe, it's more likely to be an improved breed that converts feed into milk and meat more efficiently. It will also get better veterinary care and higher-quality feed, which means it'll produce less methane.

If we can spread the improved breeds and best practices more broadly—especially crossbreeding African cows to be more productive and making higher-quality feed available and affordable—it'll reduce emissions and help poor farmers earn more money. The same is true for handling manure; rich-world farmers have access to various techniques that get rid of the manure while producing fewer emissions. As these techniques become more affordable, they'll spread to poor farmers, and we'll improve our odds of driving emissions down.

A hard-core vegan might propose another solution: *Instead of trying all these ways of reducing emissions, we should just stop raising livestock.* I can see the appeal of that argument, but I don't think it's realistic. For one thing, meat plays too important a role in human culture. In many parts of the world, even where it's scarce, eating meat is a crucial part of festivals and celebrations. In France, the gastronomic meal—including starter, meat or fish, cheese, and dessert—is officially listed as part of the country's Intangible Cultural Heritage of Humanity. According to the listing on the UNESCO website, "The gastronomic meal emphasizes togetherness, the pleasure of taste, and the balance between human beings and the products of nature."

But we can cut down on meat eating while still enjoying the taste of meat. One option is plant-based meat: plant products that have been processed in various ways to mimic the taste of meat. I've been an investor in two companies that have plant-based meat products on the market right now—Beyond Meat and Impossible Foods—so I'm biased, but I have to say that artificial meat is pretty good. When prepared just right, it's a convincing substitute for ground beef. And all of the alternatives out there are better for the environment, because they use much less land and water and are responsible for fewer emissions. You also need less grain to produce them, reducing the pressure on food crops and the use of fertilizers too. And it's a huge boon for animal welfare whenever fewer livestock are being kept in small cages.

Artificial meats come with hefty Green Premiums, however. On average, a ground-beef substitute costs 86 percent more than the real thing. But as sales for these alternatives increase, and as more of them hit the market, I'm optimistic that they'll eventually be cheaper than animal meat.

Yet the big question on artificial meat comes down to taste, not money. Although the texture of a hamburger is relatively easy to mimic with plants, it's much harder to fool people into thinking they're actually eating a steak or chicken breast. Will people like artificial meat enough to switch, and will enough people switch to make a significant difference?

We're already seeing some evidence that they will. I have to admit that even I have been surprised by how well Beyond Meat and Impossible Foods are doing—especially given their early hiccups. I attended an early demonstration by Impossible Foods at which they burned the burger so badly that it set off the smoke alarm. It's amazing to see how widely available their products are, at least around the Seattle area and the cities I visit. Beyond Meat had a very successful initial public offering in 2019. It may take another decade, but I do think that as the products get better and cheaper, people who are worried about climate change and the environment will favor them.

Another approach is akin to plant-based meat, but instead of growing plants and then processing them so they taste like beef, you grow the meat itself in a lab. It has somewhat unappealing names like "cell-based meat," "cultivated meat," and "clean meat," and there are some two dozen start-up companies working on getting it to market, though their products probably won't be on supermarket shelves until the mid-2020s.

Keep in mind that this isn't *fake* meat. Cultivated meat has all the same fat, muscles, and tendons as any animal on two or four legs. But rather than growing up on a farm, it's created in a lab. Scientists start with a few cells drawn from a living animal, let those cells multiply, and then coax them into forming all the tissues we're used to eating. All this can be done with little or no greenhouse gas emissions, aside from the electricity you need to power the labs where the process is done. The challenge with this approach is that it's very expensive, and it's not clear how much the costs can come down.

And both kinds of artificial meat face another uphill battle. At least 17 U.S. state legislatures have tried to keep these products from being labeled as "meat" in stores. One state has proposed banning their sale altogether. So even as the technology improves and the products get cheaper, we'll need to have a healthy public debate about how they're regulated, packaged, and sold.

There's one last way we can cut down on emissions from the food we eat: by wasting less of it. In Europe, industrialized parts of Asia, and sub-Saharan Africa, more than 20 percent of food is simply thrown away, allowed to rot, or otherwise wasted. In the United States, it's 40 percent. That's bad for people who don't have enough to eat, bad for the economy, and bad for the climate. When wasted food rots, it produces enough methane to cause as much warming as 3.3 billion tons of carbon dioxide each year.

The most important solution is behavior change—using more of what we have. But technology can help. For example, two companies are working on invisible, plant-based coatings that extend the life of fruits and vegetables; they're edible, and they don't affect the taste at all. Another has developed a "smart bin" that uses image recognition to track how much food is wasted in a house or business. It gives you a report on how much you threw away, along with its cost and its carbon footprint. The system may sound invasive, but giving people more information can help them make better choices.

A few years ago, I stepped into a warehouse in Dar es Salaam, Tanzania, and saw something that thrilled me: thousands of tons of synthetic fertilizer piled as high as snowdrifts. The warehouse was part of the new Yara fertilizer distribution center, which was the largest of its kind in East Africa. Walking around the warehouse, I talked to workers filling bags with tiny white pellets containing nitrogen, phosphorus, and other nutrients that would soon be nourishing crops in one of the poorest regions in the world.

It was the kind of trip I love to take. I know it sounds goofy to say this, but to me fertilizer is magical, and not just because it makes our yards and gardens prettier. Along with Norman Borlaug's semi-dwarf wheat and new varieties of corn and rice, synthetic fertilizer

Touring the Yara fertilizer distribution facility in Dar es Salaam, Tanzania, 2018. I'm having even more fun than it looks.

was a key factor in the agricultural revolution that changed the world in the 1960s and 1970s. It's been estimated that if we couldn't make synthetic fertilizer, the world's population would be 40 to 50 percent smaller than it is.

The world uses a lot of fertilizer already, and poor countries should be using more. The agricultural revolution I mentioned—often called the Green Revolution—largely bypassed Africa, where the typical farmer gets just one-fifth as much food per acre of land as an American farmer gets. That's because in poor countries most farmers don't have good enough credit to buy fertilizer, and it's more expensive than in rich countries because it has to be shipped into rural areas over poorly built roads. If we can help poor farmers raise their crop yields, they'll earn more money and have more to eat, and millions of people in some of the world's poorest countries will be able to get more food and the nutrients they need. (We'll cover this in more depth in chapter 9.)

Why is fertilizer so magical? Because it provides plants with

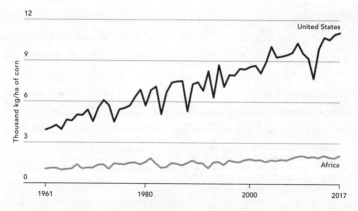

There's a huge gap in agriculture. Thanks to fertilizer and other improvements, American farmers now get more corn per unit of land than ever. But African farmers' yields have barely budged. Narrowing the gap will save lives and help people escape poverty, but without innovation it will also make climate change worse. (FAO)

essential nutrients, including phosphorus, potassium, and the one that's especially relevant to climate change: nitrogen. Nitrogen is a mixed blessing. It's closely linked to photosynthesis, the process by which plants turn sunlight into energy, so it makes all plant life—and therefore all our food—possible. But nitrogen also makes climate change much worse. To understand why, we need to talk about what it does for plants.

To grow crops, you want tons of nitrogen—way more than you would ever find in a natural setting. Adding nitrogen is how you get corn to grow 10 feet high and produce enormous quantities of seed. Oddly, most plants can't make their own nitrogen; instead, they get it from ammonia in the soil, where it's created by various microorganisms. A plant will keep growing as long as it can get nitrogen, and it'll stop once the nitrogen is all used up. That's why adding it boosts growth.

For millennia, humans fed their crops extra nitrogen by applying natural fertilizers like manure and bat guano. The big breakthrough came in 1908, when two German chemists named Fritz Haber and Carl Bosch figured out how to make ammonia from nitrogen and hydrogen in a factory. It's hard to overstate how momentous their invention was. What's now known as the Haber-Bosch process made it possible to create synthetic fertilizer, greatly expanding both the amount of food that could be grown and the range of geographies where it could be grown. It's still the main method we use to make ammonia today. In the same way that Norman Borlaug is one of the great unsung heroes of history, Haber-Bosch might be the most important invention that most people have never heard of.*

Here's the rub: Microorganisms that make nitrogen expend a lot of energy in the process. So much energy, in fact, that they've evolved to do it only when they absolutely need to—when there's no nitrogen in the soil around them. If they detect enough nitrogen, they stop producing it so they can use the energy for something else. So when we add synthetic fertilizer, the natural organisms in the soil sense the nitrogen and stop producing it on their own.

There are other downsides to synthetic fertilizer. To make it, we have to produce ammonia, a process that requires heat, which we get by burning natural gas, which produces greenhouse gases. Then, to move it from the facility where it's made to the warehouse where it's stored (like the place I visited in Tanzania) and eventually the farm where it's used, we load it on trucks that are powered by gasoline. Finally, after the fertilizer is applied to soil, much of the nitrogen that it contains never gets absorbed by the plant. In fact, worldwide, crops take up less than half the nitrogen applied to farm fields. The rest runs off into ground or surface waters, causing pollution, or escapes into the air in the form of nitrous oxide—which, you may recall, has 265 times the global-warming potential of carbon dioxide.

All told, fertilizers were responsible for roughly 1.3 billion tons of greenhouse gas emissions in 2010, and the number will probably rise

---

* Fritz Haber had a complicated history. In addition to his lifesaving work on ammonia, he pioneered the use of chlorine and other poisonous gases as chemical weapons in World War I.

to 1.7 billion tons by mid-century. Haber-Bosch giveth, and Haber-Bosch taketh away.

Unfortunately, there simply isn't a practical zero-carbon alternative for fertilizer right now. It's true that we could get rid of the emissions involved in making fertilizer by using clean electricity instead of fossil fuels to synthesize ammonia, but that's an expensive process that would raise the price of fertilizer considerably. In the United States, for example, using this process to make the nitrogen-based fertilizer urea would raise its cost by more than 20 percent.

But that's just the emissions from *making* fertilizer. We don't have any way to capture the greenhouse gases that come from *applying* it. There's no equivalent of carbon capture for nitrous oxide. That means I can't calculate a complete Green Premium for zero-carbon fertilizer—which itself is actually useful information, because it tells us that we need significant innovation in this area.

Technically, it's possible to get crops to absorb nitrogen much more efficiently than they do, if farmers have the technology to monitor their nitrogen levels very carefully and apply fertilizer in just the right amount over the course of a growing season. But that's an expensive and time-consuming process, and fertilizer is cheap (at least in rich countries). It's more economical to apply more than you need, knowing that you're at least using enough to maximize the growth of your crops.

Some companies have developed additives that are supposed to help plants take up more nitrogen so there's less to wash into groundwater or evaporate into the atmosphere. But these additives are used with only 2 percent of global fertilizers, because they don't work consistently well and manufacturers aren't investing much to improve them.

Other experts are working on different ways to solve the nitrogen problem. For example, some researchers are doing genetic work on new varieties of crops that can recruit bacteria to fix nitrogen for them. In addition, one company has developed genetically modified microbes that fix nitrogen; in effect, instead of adding nitrogen via fertilizer, you add bacteria to the soil that always produce nitrogen even when it's already present. If these approaches work, they'll dramatically reduce the need for fertilizer and all the emissions it's responsible for.

Everything you've just read about—which I'd broadly describe as agriculture—accounts for roughly 70 percent of emissions from farming, forestry, and other uses of land. If I had to sum up the other 30 percent in one word, it would be "deforestation."

According to the World Bank, the world has lost more than half a million square miles of forest cover since 1990. (That's an area bigger than South Africa or Peru, and a decline of roughly 3 percent.) There's the immediate and obvious impact of deforestation—if the trees are burned down, for example, they quickly release all the carbon dioxide they contain—but it also causes damage that's harder to see. When you take a tree out of the ground, you disturb the soil, and it turns out that there's a lot of carbon stored up in soil (in fact, there's more carbon in soil than in the atmosphere and all plant life combined). When you start removing trees, that stored carbon gets released into the atmosphere as carbon dioxide.

Deforestation would be easier to stop if it were happening for the same reasons in every place, but unfortunately that's not the case. In Brazil, for example, most of the destruction of the Amazon rain forest in the past few decades has been to clear pastureland for cattle. (Brazil's forests have shrunk by 10 percent since 1990.) And because food is a global commodity, what's consumed in one country can cause land-use changes in another. As the world eats more meat, it accelerates the deforestation in Latin America. More burgers anywhere mean fewer trees there.

And all these emissions add up fast. One study by the World Resources Institute found that if you account for land-use changes, the American-style diet is responsible for almost as many emissions as all the energy Americans use in generating electricity, manufacturing, transportation, and buildings.

But in other parts of the world, deforestation isn't about turning out more burgers and steaks. In Africa, for example, it's a matter of clearing land to grow food and fuel for the continent's growing population. Nigeria, which has had one of the highest deforestation rates in the world, has lost more than 60 percent of its forest cover since 1990, and it's one of the world's biggest exporters of charcoal, which is created by charring wood.

In Indonesia, on the other hand, forests are being cut down to make way for palm trees, which provide the palm oil you'll find in everything from movie-theater popcorn to shampoo. It's one of the main reasons why the country is the world's fourth-largest emitter of greenhouse gases.

I wish there were some breakthrough invention I could tell you about that will make the world's forests safe. There are a few things that will help, such as advanced satellite-based monitors that make it easier to spot deforestation and forest fires as they're happening and to measure the extent of the damage afterward. I'm also following some companies that are developing synthetic alternatives to palm oil so we don't have to cut down so many forests to make room for palm plantations.

But this isn't primarily a technological problem. It's a political and economic problem. People cut down trees not because people are evil; they do it when the incentives to cut down trees are stronger than the incentives to leave them alone. So we need political and economic solutions, including paying countries to maintain their forests, enforcing rules designed to protect certain areas, and making sure rural communities have different economic opportunities so they don't have to extract natural resources just to survive.

You might've heard about one forest-related solution for climate change: planting trees as a way to capture carbon dioxide from the atmosphere. Although it sounds like a simple idea—the cheapest, lowest-tech carbon capture imaginable—and it has obvious appeal for all of us who love trees, it actually opens up a very complicated subject. It needs to be studied a lot more, but for now its effect on climate change appears to be overblown.

As is so often the case in global warming, you have to consider a number of factors . . .

How much carbon dioxide can a tree absorb in its lifetime? It varies, but a good rule of thumb is 4 tons over the course of 40 years.

How long will the tree survive? If it burns down, all the carbon dioxide it was storing will be released into the atmosphere.

What would've happened if you hadn't planted that tree? If a tree would've grown there naturally, you haven't added any extra carbon absorption.

In what part of the world will you plant the tree? On balance, trees in snowy areas cause more warming than cooling, because they're darker than the snow and ice beneath them and dark things absorb more heat than light things do. On the other hand, trees in tropical forests cause more cooling than warming, because they release a lot of moisture, which becomes clouds, which reflect sunlight. Trees in the midlatitudes—between the tropics and the polar circles—are more or less a wash.

Was anything else growing in that spot? If, for example, you eliminate a soybean farm and replace it with a forest, you've reduced the total amount of soybeans available, which will drive up their price, making it more likely that someone will cut down trees somewhere else to grow soybeans. This will offset at least some of the good you do by planting your trees.

Taking all these factors into account, the math suggests you'd need somewhere around 50 acres' worth of trees, planted in tropical areas, to absorb the emissions produced by an average American in her lifetime. Multiply that by the population of the United States, and you get more than 16 billion acres, or 25 million square miles, roughly half the landmass of the world. Those trees would have to be maintained forever. And that's just for the United States—we haven't accounted for any other country's emissions.

Don't get me wrong: Trees have all kinds of benefits, both aesthetic and environmental, and we should be planting more of them. For the most part, you can get trees to grow only in places where they've already grown, so planting them could help undo the damage caused by deforestation. But there's no practical way to plant enough of them to deal with the problems caused by burning fossil fuels. The most effective tree-related strategy for climate change is to stop cutting down so many of the trees we already have.

The upshot of all this is that we'll soon need to produce 70 percent more food while simultaneously cutting down on emissions and moving toward eliminating them altogether. It'll take a lot of new ideas, including new ways to fertilize plants, raise livestock, and waste less food, and people in rich countries will need to change some habits—we'll have to eat less meat, for instance. Even if burgers run in the family.

CHAPTER 7

HOW WE GET AROUND

16 percent of 52 billion tons a year

Let's start with a quick quiz—just two questions.

    1. Which of these contains the most energy?
        A. A gallon of gasoline
        B. A stick of dynamite
        C. A hand grenade

    2. Which of these is the cheapest in the United States?
        A. A gallon of milk
        B. A gallon of orange juice
        C. A gallon of gasoline

The correct answers are A and C: gasoline. Gas contains an amazing amount of energy—you'd need to bundle 130 sticks of dynamite together to get as much energy as a single gallon of gas contains. Of course, dynamite releases all its energy at once, while gasoline burns more slowly—which is just one reason we fill up our cars with gas and not sticks of explosives.

In the United States, gasoline is also remarkably cheap, even though it may not always seem that way when it's time to stop at the gas station. In addition to milk and OJ, here are some things that it's less expensive than, gallon for gallon: Dasani bottled water, yogurt, honey, laundry detergent, maple syrup, hand sanitizer, latte from Starbucks, Red Bull energy drink, olive oil, and the famously low-cost Charles Shaw wine you can buy at Trader Joe's grocery stores. That's right—gallon for gallon, *gasoline is cheaper than Two Buck Chuck*.

As you read the rest of this chapter, keep these two facts about gasoline in mind: It packs a punch, and it's cheap.* They're a good reminder that when it comes to how much energy we get for each dollar we spend, gasoline is the gold standard. Aside from similar products like diesel and jet fuel, nothing else in our daily lives comes close to delivering as much energy per gallon at such a low cost.

The twin concepts of energy delivered per unit of fuel and per dollar spent are going to matter a lot as we look for ways to decarbonize our transportation system. As you're no doubt aware, the burning of fuels in our cars, ships, and planes emits carbon dioxide that's contributing to global warming. To get to zero, we'll need to replace those fuels with something that's just as energy dense and just as cheap.

You may be surprised that I'm bringing it up so late in this book and that transportation contributes only 16 percent of global emissions, ranking fourth behind how we make things, plug in, and grow things. I was surprised too when I learned it, and I suspect that most people are in the same boat. If you stopped some random strangers on the sidewalk and asked them what activities contribute the most to climate change, they'd probably say burning coal for electricity, driving cars, and flying planes.

The confusion is understandable: Although transportation isn't

---

* Of course, for people who rely on their cars, gasoline is more of a necessity than the other things I've listed. If you're watching your spending, you will feel the crunch of higher gas prices more than a rise in the cost of, say, olive oil, which you can always decide not to buy. But the point remains that among the things we consume on a regular basis, gasoline is relatively inexpensive.

the biggest cause of emissions worldwide, it *is* number one in the United States, and it has been for a few years now, just ahead of making electricity. We Americans drive and fly a lot.

In any case, if we're going to get to net-zero emissions, we'll have to get rid of all the greenhouse gases caused by transportation, in the United States and around the world.

How hard will that be? Pretty hard. But not impossible.

For the first 99.9 percent of human history, we managed to move around without relying on fossil fuels at all. We walked, rode animals, and put ships under sail. Then, in the early 1800s, we figured out how to run locomotives and steamboats on coal, and we never looked back. Within the century, trains were crossing entire continents and ships were moving people and products across the oceans. The gas-powered automobile came along in the late 19th century, followed in the early 20th century by the commercial air travel that would become so essential to today's global economy.

Although it's been barely 200 years since we first burned fossil fuels for transportation, we've already come to depend on them in a fundamental way. We will never give them up without a replacement that is nearly as cheap and that's just as capable of fueling long-distance travel.

Here's another challenge: We won't just need to eliminate the 8.3 billion tons of carbon we produce from transportation today; we'll need to get rid of even more than that. The Organization for Economic Cooperation and Development predicts that demand for transportation will keep growing through at least 2050—even after accounting for the fact that COVID-19 has limited travel and trade. It's aviation, trucking, and shipping—not passenger cars—that account for all the emissions growth in this sector. Maritime shipping now handles nine-tenths of the goods traded around the world by volume, producing nearly 3 percent of global emissions.

A lot of the transport emissions come from rich countries, but most of those countries hit their peak in the past decade and have actually declined somewhat since then. These days, nearly all the growth in transport-related carbon is coming from developing countries as their populations grow, get richer, and buy more cars. As usual, China is the best example—its transportation emissions have doubled over the past decade and gone up by a factor of 10 since 1990.

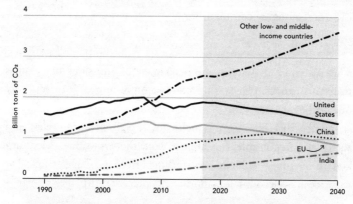

COVID-19 is slowing—but not stopping—the growth of transportation emissions. Although emissions will shrink in many places, they will grow so much in low- and middle-income countries that the overall effect will be an increase in greenhouse gases. (IEA World Energy Outlook 2020; Rhodium Group)

At the risk of sounding like a broken record, I'll make the same point about transportation that I've made about electricity, manufacturing, and agriculture: *We should be glad that more people and goods are moving around.* The ability to travel between rural areas and cities is a form of personal freedom, not to mention a matter of survival for farmers in poor countries who need to get their crops to market. International flights connect the world in ways that were unimaginable a century ago; being able to meet people from other countries helps us understand our common goals. And before modern transportation, our food choices were limited most of the year. Personally, I like grapes and enjoy eating them year-round. But I can do that only because of container ships that bring fruit from South America and that currently run on fossil fuels.

So how can we get all the benefits of travel and transportation without making the climate unlivable? Do we have all the technology we need, or do we need some innovations?

To answer those questions, we'll need to figure the Green Premiums for transportation. We'll begin by digging deeper into where these emissions are coming from.

This pie chart shows you the percentage of emissions that comes from cars, trucks, planes, ships, and so on. Our goal is to get every one of them to net zero.

Notice that passenger vehicles (cars, SUVs, motorcycles, and such) are responsible for almost half the emissions. Medium- and heavy-duty vehicles—everything from garbage trucks to 18-wheelers—account for another 30 percent. Airplanes add in a

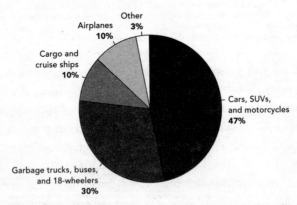

Cars aren't the only culprit. Passenger vehicles are responsible for nearly half of all transportation-related emissions. (International Council on Clean Transportation)

tenth of all emissions, as do container ships and other marine vessels, with trains accounting for the last bit.*

Let's take these one at a time, starting with the biggest slice of the pie—passenger cars—and look at our current options for getting rid of emissions.

Passenger cars. There are about a billion cars on the road around the world. In 2018 alone, we added roughly 24 million passenger cars, after accounting for the ones that got retired. Because burning gasoline inevitably releases greenhouse gases, we need an alternative—either fuels made from carbon that's already in the air rather than the carbon that's in fossil fuels, or some other form of energy altogether.

Let's take the second option first. Fortunately, we do indeed have another form of energy that—although far from perfect—has already been proven to work. In fact, cars that use it are probably being sold at an auto dealer near you right now.

Today you can buy an all-electric car from more than half the alphabet: Audi, BMW, Chevrolet, Citroën, Fiat, Ford, Honda, Hyundai, Jaguar, Kia, Mercedes-Benz, Nissan, Peugeot, Porsche, Renault, Smart, Tesla, Volkswagen, and others too numerous to mention, including manufacturers in China and India. I own an electric vehicle, and I love it.

Although EVs used to be far more expensive than their gas-burning counterparts, and they're still the pricier option today, the difference has come down dramatically in recent years. That's largely due to a huge drop in the cost of batteries—an 87 percent decrease since 2010—as well as various tax credits and government commitments to get more zero-emissions cars on the road. But EVs still come with a modest Green Premium.

---

* As a reminder, I'm counting only emissions from the fuel that various vehicles burn. The emissions from manufacturing them—making the steel and plastic, running the factories, and so on—are counted under "How we make things" and covered in chapter 5.

For example, consider two cars, both produced by Chevrolet: the gas-powered Malibu and the all-electric Bolt EV.

Malibu
Starting at $23,400

MPG: 29 city / 36 hwy
Cargo: 15.7 cubic feet
Horsepower: 250

Bolt EV
Starting at $31,500

Range: 259 miles
Cargo: 57 cubic feet
Horsepower: 200

Chevy versus Chevy. The gas-powered Malibu and the all-electric Bolt EV. (Chevrolet)

Their features are roughly comparable when it comes to engine power, the amount of space for passengers, and so on. The Bolt costs $8,100 more (before any tax incentives that might make it cheaper), but you can't figure the Green Premium using only the purchase price of the car. What matters isn't just the cost of buying the car but the overall cost of buying *and owning* the car. You have to account for the fact that EVs need less maintenance, for example, and run on electricity instead of gas. On the other hand, because EVs are more expensive, you'll pay more for auto insurance.

When you account for all these differences and look at the total cost of ownership, the Bolt will cost 10 cents more per mile driven than the Malibu.

What does 10 cents a mile mean? If you drive 12,000 miles a year, that's an annual premium of $1,200—hardly negligible, but low enough to make EVs a reasonable consideration for many car buyers.

And that's a national average in the United States. The Green Premium will be different in other countries—the main factor being the difference between the cost of electricity and the cost of gasoline. (Cheaper electricity or more expensive gasoline will make the Green Premium smaller.) In some parts of Europe, gas prices are so high that the Green Premium for EVs has already reached zero. Even in the United States, as battery prices continue to drop, I predict that the premium for most cars will be zero by 2030.

That's great news, and we should get lots of EVs on the road as they become even more affordable. (I'll say more about how we can do this at the end of this chapter.) But even in 2030, there will be some drawbacks to EVs versus a gas-powered car.

One is that gasoline prices vary a lot, and EVs are the cheaper option only when gas prices are above a certain level. At one point in May 2020, the average price of gas in the United States had dropped to $1.77 per gallon; when gas is that cheap, EVs can't compete—the batteries are simply too expensive. With the price of today's batteries, EV owners save money only if gas costs more than around $3 per gallon.

Another drawback is that it takes an hour or more to fully charge an EV, yet you can gas up your car in less than five minutes. In addition, using them to avoid carbon emissions works only if we're generating electricity from zero-carbon sources. This is another reason why the breakthroughs I mentioned in chapter 4 are so important. If we get our power from coal and then charge up our electric cars with coal-fired electricity, we'll just be swapping one fossil fuel for another.

Plus, it'll take time to get all our gas-burning cars off the road. On average, after a car rolls off the assembly line, it runs for more than 13 years before reaching its final resting place in the junkyard. This long life cycle means that if we wanted to have every passenger car in America running on electricity by 2050, EVs would need to be nearly 100 percent of auto sales within the next 15 years. Today they're about 3 percent.

As I mentioned, another way to get to zero is to switch to alternative liquid fuels that use carbon that was already in the atmosphere. When you burn these fuels, you're not adding extra carbon to the air—you're just returning the same carbon to where it was when the fuel was made.

When you see the phrase "alternative fuels," you might think about ethanol, a biofuel that's usually made from corn, sugarcane, or beet sugar. If you're in the United States, you're probably running your car on some of this biofuel already—most gasoline sold in America contains 10 percent ethanol, virtually all of it made from corn. There are cars in Brazil that run on 100 percent ethanol made from sugarcane. Few other countries use any at all.

Here's the problem: Corn-based ethanol isn't zero carbon, and depending on how it's made, it may not even be low carbon. Growing the crops requires fertilizer. The refining process, when the plants get turned into fuel, produces emissions too. And growing crops for fuel takes up land that might otherwise be used for growing food—possibly forcing farmers to cut down forests so they have someplace to grow food crops.

Alternative fuels are not a lost cause, though. There are advanced, second-generation biofuels that don't have the problems of conventional biofuels. They can be made from plants that aren't grown for food—unless you're a big fan of switchgrass salad—or from farming residue (such as cornstalks), by-products left over from making paper, and even food and yard waste. Because they're not food crops, they need little or no fertilizer, and they don't have to be grown on farmland that could otherwise be dedicated to food for people or animals.

Some advanced biofuels will be what experts call "drop-in" fuels—meaning you can use them in (or "drop them into") a conventional engine without modifying it. One more benefit: We can transport them using the tankers, pipelines, and other infrastructure we've already spent billions to build and maintain.

I'm optimistic about biofuels, but it's a tough field. I had a personal experience that shows just how hard it is to make a breakthrough. A few years ago, I learned about a U.S. company that had a proprietary process for converting biomass, such as trees, into fuels. I went to visit its plant and was impressed by what I saw, and after doing due diligence, I invested $50 million in the company. But its technology just didn't work well enough—various technical challenges meant the plant couldn't produce at nearly the volume it needed to be economical—and the plant I visited eventually shut down. It was a $50 million dead end, but I'm not sorry I did it. We need to be exploring lots of ideas, even knowing that many of them will fail.

Unfortunately, research on advanced biofuels is still underfunded, and they're not ready to be deployed at the scale we need for decarbonizing our transportation system. As a result, using them to replace gasoline would be quite expensive. Experts disagree on the exact cost of these and other clean fuels, and there's a range of estimates out there, so I'll use average costs from several different studies.

Green Premium to replace gasoline with advanced biofuels

Fuel type	Retail price per gallon	Zero-carbon option per gallon	Green Premium
Gasoline	$2.43	$5.00 (advanced biofuels)	106%

NOTE: Retail prices in this and subsequent charts are the average in the United States from 2015 to 2018. Zero-carbon options reflect current estimated prices.

Biofuels get their energy from plants, but that's not the only way to create alternative fuels. We can also use zero-carbon electricity to combine the hydrogen in water with the carbon in carbon dioxide, resulting in hydrocarbon fuels. Because you use electricity in the process, these fuels are sometimes called electrofuels, and they have a lot of advantages. They're drop-in fuels, and because they're made using carbon dioxide captured from the atmosphere, burning them doesn't add to overall emissions.

But electrofuels also have a downside: They're very expensive. You need hydrogen to make them, and as I mentioned in chapter 4, it costs a lot to make hydrogen without emitting carbon. In addition, you need to make them using clean electricity—otherwise, there's no point—and we don't yet have enough cheap, clean electricity in our power grid to use it economically for making fuel. It all adds up to a high Green Premium for electrofuels:

Green Premiums to replace gasoline with zero-carbon alternatives

Fuel type	Retail price per gallon	Zero-carbon option per gallon	Green Premium
Gasoline	$2.43	$5.00 (advanced biofuels)	106%
Gasoline	$2.43	$8.20 (electrofuels)	237%

What does that mean for the average family? In the United States, a typical household spends around $2,000 a year on gasoline. So if the price doubles, that's an extra $2,000 premium, and if it triples, that's an extra $4,000 for every conventional passenger car on the road in America.

Garbage trucks, buses, and 18-wheelers. Unfortunately, batteries are a less practical option when it comes to long-distance buses and trucks. The bigger the vehicle you want to move, and the farther you want to drive it without recharging, the harder it'll be to use electricity to power your engine. That's because batteries are heavy, they can store only a limited amount of energy, and they can deliver only a certain amount of that energy to the engine at one time. (It takes a more powerful engine—one with more batteries—to run a heavy truck than a light hatchback.)

Medium-duty vehicles, like garbage trucks and city buses, are generally lightweight enough that electricity is a viable option for them. They also have the advantage of running relatively short routes and parking in the same place every night, so it's easy to set up charging stations for them. The city of Shenzhen, China—home to 12 million people—has electrified its entire fleet of more than 16,000 buses and nearly two-thirds of its taxis. With the volume of electric buses being sold in China, I think the Green Premium for buses will reach zero within a decade, which means that most cities in the world will be able to shift their fleets.

Shenzhen, China, electrified its fleet of 16,000 buses.

But if you want to add more distance and power—for example, if you're trying to run an 18-wheeler loaded with cargo on a cross-country trip, rather than a school bus full of students on a route around the neighborhood—you'll need to carry many more batteries. And as you add batteries, you also add weight. A *lot* of weight.

Pound for pound, the best lithium-ion battery available today packs 35 times less energy than gasoline. In other words, to get the same amount of energy as a gallon of gas, you'll need batteries that weigh 35 times more than the gas.

Here's what that means in practical terms. According to a 2017 study by two mechanical engineers at Carnegie Mellon University, an electric cargo truck capable of going 600 miles on a single charge would need so many batteries that it would have to carry 25 percent less cargo. And a truck with a 900-mile range is out of the question:

It would need so many batteries that it could hardly carry any cargo at all.

Keep in mind that a typical truck running on diesel can go more than 1,000 miles without refueling. So to electrify America's fleet of trucks, freight companies would have to shift to vehicles that carry less cargo, stop to recharge far more often, spend hours of time recharging, and somehow travel long stretches of highway where there are no recharging stations. It's just not going to happen anytime soon. Although electricity is a good option when you need to cover short distances, it's not a practical solution for heavy, long-haul trucks.

Because we can't electrify our cargo trucks, the only solutions available today are electrofuels and advanced biofuels. Unfortunately, they have dramatic Green Premiums too. Let's add them to the chart:

Green Premiums to replace diesel with zero-carbon alternatives

Fuel type	Retail price per gallon	Zero-carbon option per gallon	Green Premium
Diesel	$2.71	$5.50 (advanced biofuels)	103%
Diesel	$2.71	$9.05 (electrofuels)	234%

Ships and planes. Not long ago, my friend Warren Buffett and I were talking about how the world might decarbonize airplanes. Warren asked, "Why can't we run a jumbo jet on batteries?" He already knew that when a jet takes off, the fuel it's carrying accounts for 20 to 40 percent of its weight. So when I told him this startling fact—that you'd need 35 times more batteries by weight to get the same energy as jet fuel—he understood immediately. The more power you need, the heavier your plane gets. At some point, it's so heavy that it can't get off the ground. Warren smiled, nodded, and just said, "Ah."

When you're trying to power something as heavy as a container ship or jetliner, the rule of thumb I mentioned earlier—*the bigger the vehicle you want to move, and the farther you want to drive it without recharging, the harder it'll be to use electricity as your power source*—becomes a law. Barring some unlikely breakthrough, batteries will never be light and powerful enough to move planes and ships anything more than short distances.

Consider where the state of the art is today. The best all-electric plane on the market can carry two passengers, reach a top speed of 210 miles per hour, and fly for three hours before recharging.* Meanwhile, a mid-capacity Boeing 787 can carry 296 passengers, reach up to 650 miles an hour, and fly for nearly 20 hours before stopping for fuel. In other words, a fossil-fuel-powered jetliner can fly more than three times as fast, for six times as long, and carry nearly 150 times as many people as the best electric plane on the market.

Batteries are getting better, but it's hard to see how they'll ever close this gap. If we're lucky, they may become up to three times as energy dense as they are now, in which case they would still be 12 times less energy dense than gas or jet fuel. Our best bet is to replace jet fuel with electrofuels and advanced biofuels, but look at the hefty premiums that come with them:

Green Premiums to replace jet fuel with zero-carbon alternatives

Fuel type	Retail price per gallon	Zero-carbon option per gallon	Green Premium
Jet fuel	$2.22	$5.35 (advanced biofuels)	141%
Jet fuel	$2.22	$8.80 (electrofuels)	296%

The same goes for cargo ships. The best conventional container ships can carry 200 times more cargo than either of the two electric ships now in operation, and they can run routes that are 400 times

---

* Air speed is usually measured in knots, but most people (including me) don't know how much a knot is. In any case, knots are pretty close to miles per hour.

longer. Those are major advantages for ships that need to cross entire oceans.

Given how important container ships have become in the global economy, I don't think it will ever be financially viable to try to run them on anything other than liquid fuels. Making the switch to alternatives would do us a lot of good; because shipping alone accounts for 3 percent of all emissions, using clean fuels would give us a meaningful reduction. Unfortunately, the fuel that container ships run on—it's called bunker fuel—is dirt cheap, because it's made from the dregs of the oil-refining process. Since their current fuel is so inexpensive, the Green Premium for ships is very high:

Green Premiums to replace bunker fuel with zero-carbon alternatives

Fuel type	Retail price per gallon	Zero-carbon option per gallon	Green Premium
Bunker fuel	$1.29	$5.50 (advanced biofuels)	326%
Bunker fuel	$1.29	$9.05 (electrofuels)	601%

To sum up, here are all the Green Premiums from this chapter:

Green Premiums to replace current fuels with zero-carbon alternatives

Fuel type	Retail price per gallon	Zero-carbon option per gallon	Green Premium
Gasoline	$2.43	$5.00 (advanced biofuels)	106%
Gasoline	$2.43	$8.20 (electrofuels)	237%
Diesel	$2.71	$5.50 (advanced biofuels)	103%
Diesel	$2.71	$9.05 (electrofuels)	234%
Jet fuel	$2.22	$5.35 (advanced biofuels)	141%
Jet fuel	$2.22	$8.80 (electrofuels)	296%
Bunker fuel	$1.29	$5.50 (advanced biofuels)	326%
Bunker fuel	$1.29	$9.05 (electrofuels)	601%

Would most people be willing to accept these increases? It's not clear. But consider that the last time the United States raised the federal gas tax—imposed any increase at all—was more than a quarter century ago, in 1993. I don't think Americans are eager to pay more for gas.

There are four ways to cut down on emissions from transportation. One is to do less of it—less driving, flying, and shipping. We should encourage more alternative modes, like walking, biking, and carpooling, and it's great that some cities are using smart urban plans to do just that.

Another way to cut down on emissions is to use fewer carbon-intensive materials in making cars to begin with—although that wouldn't affect the fuel-based emissions we've covered in this chapter. As I mentioned in chapter 5, every car is made from materials like steel and plastics that can't be manufactured without emitting greenhouse gases. The less of these materials we need in our cars, the lower their carbon footprint will be.

The third way to cut down on emissions is to use fuels more efficiently. This subject gets a lot of attention from lawmakers and the press, at least as it pertains to passenger cars and trucks; most major economies have fuel efficiency standards for those vehicles, and they've made a big difference by forcing car companies to fund the advanced engineering of more efficient engines.

But the standards don't go far enough. For example, there are suggested emissions standards for international shipping and aviation, but they're almost unenforceable. Which country's jurisdiction would cover carbon emissions from a container ship in the middle of the Atlantic Ocean?

Besides, although making and using more efficient vehicles are important steps in the right direction, they won't get us to zero. Even if you're burning less gasoline, you're still burning gasoline.

That brings me to the fourth—and most effective—way we can move toward zero emissions from transportation: switching to electric vehicles and alternative fuels. As I've argued in this chapter, both options currently carry a Green Premium to one degree or another. Let's look at ways to reduce it.

How to Lower the Green Premium

For passenger cars, the Green Premium is on the way down and will eventually shrink to zero. It is true that as higher-mileage cars and EVs replace today's vehicles, the revenue from gas taxes will go down, which could reduce the funding that's available for building and maintaining roads. States can replace the lost revenue by charging EV owners an extra fee when they renew their license plates—19 states are doing this as I write this chapter—though it means it'll take a year or two longer for EVs to be as cheap as gas-fueled cars.

EVs are driving into another headwind too: America's love for big, gas-guzzling trucks. In 2021, we bought more than 3 million cars and 12 million trucks and SUVs. All but 3 percent of these vehicles run on gasoline.

To turn things around, we'll need some inventive government policies. We can speed up the transition by adopting policies that encourage people to buy EVs and creating a network of charging stations so they're more practical to own. Nationwide commitments can help drive up the supply of cars and drive down their cost; China, India, and several countries in Europe have all announced goals to phase out fossil-fueled vehicles—mostly passenger cars—over the coming decades. California has committed to buying only electric buses by 2029 and to banning the sale of gas-powered cars by 2035.

Next, to run all these EVs we hope to have on the road, we'll need a lot of clean electricity—one more reason why it's so important to deploy renewable sources and pursue the breakthroughs in generation and storage that I mentioned in chapter 4.

We should also be exploring nuclear-powered container ships. The risks here are real (for example, you have to make sure the nuclear fuel doesn't get released if the ship sinks), but many of the technical challenges have already been solved. After all, military submarines and aircraft carriers run on nuclear power already.

Finally, we need a massive effort to explore all the ways we can make advanced biofuels and cheap electrofuels. Companies and researchers are exploring several different pathways—for example, new ways to make hydrogen using electricity, or using solar power, or using microbes that naturally produce hydrogen as a by-product. The more we explore, the more opportunities we'll create for breakthroughs.

It's rare that you can boil the solution for such a complex subject down into a single sentence. But with transportation, the zero-carbon future is basically this: Use electricity to run all the vehicles we can, and get cheap alternative fuels for the rest.

In the first group are passenger cars and trucks, light and medium trucks, and buses. In the second group are long-distance trucks, trains, airplanes, and container ships. As for cost, electric passenger cars will soon be no more expensive to own than gas-powered ones, which is great; but alternative fuels are still quite expensive, which isn't great. We need innovation to bring those prices down.

This chapter has covered how we move people and goods around from place to place. Next we'll talk about the places we're headed to—our homes, offices, and schools—and what it takes to keep them livable in a warmer world.

CHAPTER 8

HOW WE KEEP COOL
AND STAY WARM

7 percent of 52 billion tons a year

I never thought I'd find something to like about malaria. It kills 400,000 people a year, most of them children, and the Gates Foundation is part of a global push to eradicate it. So I was surprised when I learned a while back that there is actually one nice thing you can say about malaria: It helped give us air-conditioning.

Humans have been trying to beat the heat for millennia. Buildings in ancient Persia were equipped with wind catchers, or *badgirs,* which helped keep the air moving and the temperature cool. But the first known machine to produce cold air was created in the 1840s by John Gorrie, a physician in Florida who thought cooler temperatures would help his patients recover from malaria.

Back then, it was widely believed that malaria was caused not by a parasite, as we now know it is, but by bad air (hence the name, *mal-aria*). Gorrie set up a device that cooled off his sick ward by moving air over a big block of ice suspended from the ceiling. But the machine went through ice quickly, and ice was expensive because it had to be shipped in from the north, so Gorrie designed a machine to make it himself. He eventually received a patent for his ice maker, and he left medicine to try to market his invention. Unfortunately, his business plans didn't pan out. After a series of misfortunes, Gorrie died penniless in 1855.

Still, the idea took hold. The next big air-conditioning advance was made in 1902 by an engineer named Willis Carrier, when his employer sent him to a print shop in New York to figure out how to keep the pages of magazines from wrinkling as they came off the printing press. Realizing that the wrinkles were caused by high humidity levels, Carrier designed a machine that lowered the humidity while also decreasing the temperature in the room. He didn't know it yet, but he had given birth to the air-conditioning industry.

Barely more than a century after the first A/C unit was installed in a private residence, 90 percent of American households now have some type of air conditioner. If you've ever enjoyed a game or concert in a domed stadium, you can thank air-conditioning. And it's hard to imagine that places like Florida and Arizona would be attractive destinations for retirees today without it.

Air-conditioning is no longer simply a pleasant luxury that makes summer days bearable; the modern economy depends on it. To take just one example: The server farms containing thousands of computers that make today's computing advances possible (including the ones that run the cloud services where you store music and photos) generate huge amounts of heat. If they didn't stay cool, the servers would melt down.

If you live in a typical American home, your air conditioner is the biggest consumer of electricity you own—more than your lights, refrigerator, and computer combined.* I counted electricity emissions in chapter 4, but I'm mentioning them again here because space cooling is such a key contributor now and in the future. Also, although A/C units demand the most *electricity*, they're not the

---

* Electricity accounts for 99 percent of the energy used for space cooling around the world. Most of the other 1 percent is accounted for by natural-gas-fueled chillers for air-conditioning. Natural-gas-fired air-conditioning systems are available for single-family homes, but it's such a small percentage of the market that the Energy Information Administration doesn't even collect data on it.

largest consumers of *energy* in American homes and businesses. That honor goes to our furnaces and water heaters. (This is also true in Europe and many other regions.) I'll get to heating air and water in the next section.

Americans are not alone in liking—and needing—cool air. Worldwide, there are 1.6 billion A/C units in use, but they're not evenly distributed. In rich countries like the United States, 90 percent or more of households have air-conditioning, while in the hottest countries in the world less than 10 percent do.

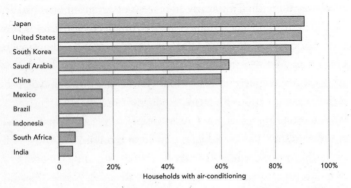

A/C is on the way. In some countries, most houses have air-conditioning, but in others it is much less common. In the coming decades, the countries at the bottom of this chart will be getting both hotter and richer, which means they'll be buying and running more A/C units. (IEA)

That means we'll be adding many more units as the population grows and gets richer and as heat waves become more severe and more frequent. China added 350 million units between 2007 and 2017 and is now the largest market in the world for air conditioners. Worldwide, sales rose 15 percent in 2018 alone, with much of the growth coming from four countries where temperatures get especially high: Brazil, India, Indonesia, and Mexico. By 2050, there will be more than 5 billion A/C units in operation around the world.

Ironically, the very thing we'll be doing to survive in a warmer climate—running air conditioners—could make climate change worse. After all, air conditioners run on electricity, so as we install more of them, we'll need more electricity to run them. In fact, the International Energy Agency projects that worldwide electricity demand for cooling will triple by 2050. At that point, air conditioners will consume as much electricity as all of China and India do now.

That will be good for people who suffer in heat waves, but bad for the climate, because in most parts of the world generating power is still a carbon-intensive process. That's why all the electricity used by buildings—for air-conditioning as well as lights, computers, and so on—is responsible for nearly 14 percent of all greenhouse gases.

The fact that air-conditioning relies so much on electricity makes it easy to calculate the Green Premium for cool air. To decarbonize our air conditioners, we need to decarbonize our power grids. This is another reason why we need breakthroughs in generating and storing electricity like the ones I described in chapter 4; otherwise, emissions will keep going up, and we'll be stuck in a vicious cycle, making our homes and offices progressively cooler while making the climate progressively hotter.

Fortunately, we don't have to wait for those breakthroughs. We can take action now to reduce the amount of electricity needed for air conditioners, and therefore lower the emissions caused by staying cool. And there's no technical barrier to doing this. Most people simply don't buy the most energy-conscious air conditioners on the market. According to the IEA, the typical A/C unit sold today is only half as efficient as what's widely available and only a third as efficient as the best models.

Mostly that's because consumers can't get all the information they need when they're picking an air conditioner. For example, a less efficient unit might be cheaper to buy but more expensive to own in the long run, because it uses more electricity. Yet if the units aren't labeled clearly, you might not have any way to know that when you're shopping around. (Such labels are required in the United States but not globally.) Plus, many countries don't set minimum standards for the efficiency of air conditioners. The IEA found that simply by creating policies to fix problems like these, the world can double the average efficiency of A/C units and reduce the growth of energy demand for cooling at mid-century by 45 percent.

Unfortunately, their demand for electricity isn't the only thing that makes air conditioners a problem. They also contain refrigerants—known as F-gases, because they contain fluorine— that leak out little by little over time when the unit ages and breaks down, as you've no doubt noticed if you've ever had to replace the coolant in your car's air conditioner. F-gases are extremely powerful contributors to climate change: Over the course of a century, they cause thousands of times more warming than an equivalent amount of carbon dioxide. If you don't hear much about them, it's because they're not a huge percentage of greenhouse gases; in the United States, they represent about 3 percent of emissions.

Yet F-gases haven't gone unnoticed. In 2016, representatives from 197 countries committed to reducing the production and use of certain F-gases by more than 80 percent by 2045—a commitment they could make because various companies are developing new approaches to air-conditioning that replace F-gases with less harmful coolants. These ideas are in the early stages of development, far too early to put a price tag on them, but they're good examples of the kind of innovation we'll need to keep cool without making the world any warmer.

In a book about global warming, it may seem strange to discuss staying warm. Why turn up the thermostat when it's already hot outside? For one thing, when we talk about heat, we're not just talking about making the air warmer; we also have to heat water for everything from showers and dishwashers to industrial processes. But more to the point, winter isn't going away. Even as global temperatures go up overall, it's still going to freeze and snow in many places around the world. And winters are especially hard for anyone who relies on renewables. In Germany, for example, during the winter the amount of solar power available can drop by as much as a factor of nine, and there are also periods with no wind. But you still need electricity; without it, people will freeze to death in their own homes.

Together, furnaces and water heaters account for a third of all emissions that come from the world's buildings. And unlike lights and A/C units, most of them run on fossil fuels, not electricity. (Whether you use natural gas, heating oil, or propane depends largely on where you live.) This means we can't decarbonize hot water and air simply by cleaning up our electric grid. We need to get heat from something other than oil and gas.

The path to zero carbon for heating actually looks a lot like the path for passenger cars: (1) electrify what we can, getting rid of natural gas water heaters and furnaces, and (2) develop clean fuels to do everything else.

The good news is that step 1 can actually carry a negative Green Premium. Unlike electric cars, which are more expensive to own than their gas-powered counterparts, all-electric heating and cooling lets you save money. And that's true whether you're building new construction from scratch or retrofitting an older home. In most locations, your overall costs will go down if you get rid of an electric air conditioner and gas (or oil) furnace and replace both with an electric heat pump.

The idea of a heat pump can seem odd the first time you hear it. Although it's easy to imagine pumping water or air, how on earth would you pump heat?

Heat pumps take advantage of the fact that gases and liquids change temperature as they expand and contract. The pumps work by moving some coolant through a closed loop of pipes, using a compressor and special valves to change the pressure along the way so that the coolant absorbs heat from one place and gets rid of it somewhere else. In the winter, you move heat from outdoors into your home (this is possible in all but the very coldest climates); in the summer, you do the opposite, pumping heat from inside your house to the outdoors.

This isn't as mysterious as it may sound. You already have a heat pump in your home, and it's probably operating right now. It's called a refrigerator. The warm air that you feel blowing from the bottom of your fridge is what carries the heat away from your food and keeps it cool.

How much money can a heat pump save you? It varies from city to city, depending on how harsh the winter is, how much electricity and natural gas cost, and other factors. Here are a few examples of the savings on new construction in cities around the United States, including the cost of installing a heat pump and operating it for 15 years:

Green Premium for installing an air-sourced heat pump in selected U.S. cities

City	Cost of natural gas furnace & electric A/C	Cost of air-sourced heat pump	Green Premium
Providence, RI	$12,667	$9,912	-22%
Chicago, IL	$12,583	$10,527	-16%
Houston, TX	$11,075	$8,074	-27%
Oakland, CA	$10,660	$8,240	-23%

You won't save as much if you're retrofitting an existing home, but switching to a heat pump is still less expensive in most cities. In Houston, for example, doing this will save you 17 percent. In Chicago, your costs will actually go up 6 percent, because natural gas there is unusually cheap. And in some older homes it's simply not practical to find space for new equipment, so you might not be able to upgrade at all.

Still, these negative Green Premiums raise an obvious question: If heat pumps are such a great deal, why are they in only 11 percent of American homes?

Partly it's because we replace our furnaces only every decade or so, and most people don't have enough extra cash on hand to simply replace a perfectly good furnace with a heat pump.

But there's another explanation as well: outdated government policies. Since the energy crisis of the 1970s, we've been trying to cut down on energy use, and so state governments created various incentives to favor natural gas furnaces and water heaters over less efficient electric ones. Some modified their building codes to make it harder for homeowners to replace their gas appliances with electric alternatives. Many of these policies that prize efficiency over emissions are still on the books, restricting your ability to lower your emissions by swapping out a gas-burning furnace for an electric heat pump—even if doing this would save you money.

This is frustrating in that familiar "regulations really can be dumb" way. But if you look at it from a different angle, it's good news. It means we don't need some additional technological breakthrough to reduce our emissions in this area, beyond decarbonizing our power grid. The electric option already exists, it's widely available, and it isn't merely price competitive—it's actually cheaper. We just need to make sure our government policies keep up with the times.

Unfortunately, although it's technically *possible* to zero out heating emissions by going electric, it won't happen quickly. Even if we fixed the self-defeating regulations I mentioned, it's not realistic to think we'll simply rip out all our gas furnaces and water heaters and replace them with electric ones overnight, any more than we're suddenly going to run the world's fleet of passenger cars on electricity. Given how long today's furnaces last, if we had a goal of getting rid of all the gas-powered ones by mid-century, we'd have to stop selling them by 2035. Today around half of all furnaces sold in the United States run on gas; worldwide, fossil fuels provide six times more energy for heating than electricity does.

To me, that's another argument for why we need advanced biofuels and electrofuels like the ones I mentioned in chapter 7—ones that can be run in the furnaces and water heaters we have today, without modification, and that don't add more carbon to the atmosphere. But right now, both options carry a hefty Green Premium:

Green Premiums to replace current heating fuels with zero-carbon alternatives

Fuel type	Current retail price	Zero-carbon option	Green Premium
Heating oil (per gallon)	$2.71	$5.50 (advanced biofuels)	103%
Heating oil (per gallon)	$2.71	$9.05 (electrofuels)	234%
Natural gas (per therm)	$1.01	$2.45 (advanced biofuels)	142%
Natural gas (per therm)	$1.01	$5.30 (electrofuels)	425%

NOTE: Retail price per gallon is the average in the United States from 2015 to 2018. Zero-carbon is current estimated price.

Let's look at what these premiums would mean for a typical U.S. family. If they heat their home with fuel oil, they're going to pay $1,300 more if they want to use advanced biofuels, and more than $3,200 extra if they choose electrofuels. If their home is heated with natural gas, switching to advanced biofuels would add $840 to their bill each winter. Switching to electrofuels would add nearly $2,600 each winter.

Clearly we need to drive down the price of these alternative fuels, as I argued in chapter 7. And there are other steps we can take to decarbonize our heating systems:

Electrify as much as we can, getting rid of gas-powered furnaces and water heaters and replacing them with electric heat pumps. In some regions, governments will have to update their policies to allow—and encourage—these upgrades.

Decarbonize the power grid by deploying today's clean sources where they make sense and investing in breakthroughs for generating, storing, and transmitting power.

Use energy more efficiently. This may seem like a contradiction, because just a few paragraphs ago I complained about policies that prize higher efficiency over lower emissions. The truth is, we need both.

The world is undergoing a huge construction boom. To accommodate a growing urban population, we'll add 2.5 trillion square feet of buildings by 2060—the equivalent, as I mentioned in chapter 2, of putting up another New York City every month for 40 years. It's a fair bet that many of these buildings will not be designed to conserve energy and that they'll be around, using energy inefficiently, for several decades.

The good news is that we know how to build green buildings—as long as we're willing to pay a Green Premium. An extreme example is Seattle's Bullitt Center, which lays claim to being one of the greenest commercial buildings in the world. The Bullitt Center was designed to naturally stay warm in the winter and cool in the summer, reducing the need for heating and air-conditioning, and features other energy-saving technologies such as a superefficient elevator. At times, it can generate 60 percent more energy than it consumes, thanks to solar panels on its roof, although it's still plugged into the city's electric grid and draws power at night and during especially cloudy stretches. Which we have plenty of here in Seattle.

Although many of the Bullitt Center's technologies are currently too expensive for widespread use (which is why it remains one of the world's greenest buildings seven years after it opened), we can still make homes and offices more efficient at a low cost. They can be designed with what developers call a supertight envelope (not much air leaking in or out), good insulation, triple-glazed windows, and efficient doors. I'm also intrigued by windows that use so-called smart glass, which automatically turns darker when the room needs to be cooler and lighter when it needs to be warmer. New building regulations can help promote these energy-saving ideas, which will

The Bullitt Center in Seattle is one of the greenest commercial buildings in the world.

expand the market and drive down their cost. We can make a lot of buildings more energy efficient, even if they can't all be as efficient as the Bullitt Center.

We've now covered all five major sources of greenhouse gas emissions: how we plug in, make things, grow things, get around, and keep cool and warm. I hope three things are clear by now:

1. The problem is extremely complex, touching on almost every human activity.
2. We have some tools at hand that we should be deploying now to reduce emissions.
3. But we don't have all the tools we need. We need to drive down the Green Premiums in every sector, which means we've got a lot of inventing to do.

In chapters 10 through 12, I'll suggest the specific steps that I think will give us the best chance of developing and deploying the tools we need. But first, I want to confront a question that keeps me up at night. So far, this book has been exclusively about how to lower emissions and keep the temperature from becoming unbearable. What can we do about the climate changes that are already happening? And, in particular, how can we help the world's poorest, who have the most to lose but did the least to cause the problem?

CHAPTER 9

ADAPTING TO A WARMER WORLD

've been making the case that we need to get to zero emissions and that we're going to need a lot of innovation to do it. But innovation doesn't happen overnight, and it will take decades for the green products I've been telling you about to reach a big enough scale to make a significant difference.

In the meantime, people all over the world, at every income level, are already being affected in one way or another by climate change. Just about everyone who's alive now will have to adapt to a warmer world. As sea levels and floodplains change, we'll need to rethink where we put homes and businesses. We'll need to shore up power grids, seaports, and bridges. We'll need to plant more mangrove forests (stay tuned if you don't know what a mangrove is) and improve our early-warning systems for storms.

I'll return to those projects later in this chapter. But now I want to tell you about the people I think of first when I think about who will suffer the most from a climate disaster and who deserve the most help adapting to it. They don't have much in the way of power grids, seaports, or bridges to worry about. They're the low-income people I meet through my work on global health and development, and for them climate change could have the worst consequences of all. And their stories capture the complexity of fighting poverty and climate change at the same time.

For example, in 2009, I met the Talam family—Laban, Miriam, and their three children—when I was in Kenya to learn about the lives of farmers with less than four acres of land (or, as they're known in development parlance, smallholder farmers). I visited their farm a few miles down a dirt road outside Eldoret, one of the fastest-growing cities in Kenya. The Talams didn't have much, just a few round mud huts with thatched roofs and an animal pen, and their farm covered about two acres, smaller than a baseball field. But what was happening on this small plot of land was drawing hundreds of farmers from miles around to learn what the owners were doing and how they could do it themselves.

I got to visit Miriam and Laban Talam on their farm in Kabiyet, Kenya, in 2009. They have an amazing story of success, but climate change could undo all the progress they've made.

Laban and Miriam greeted me at their front gate and started telling me their story. Two years before, they had been smallholder farmers practicing subsistence farming. Like most of their neighbors, they had been desperately poor. They grew corn (in Kenya, as in many places around the world, it's known as maize) and other vegetables, some to eat themselves and the rest to sell at market. Laban would work odd jobs to make ends meet. To earn more income, he had bought a cow, which the couple would milk twice a day: They'd sell the morning milk to a local trader to get a small amount of cash, and they'd save the evening milk for themselves and their children. In all, the cow would produce three liters of milk per day; that's less than a gallon each day to sell and split among a family of five.

By the time I met them, life for the Talams had improved dramatically. They now had four cows, which were producing 26 liters of milk every day. They sold 20 liters a day and kept six liters for themselves. Their cows earned them nearly $4 per day, which in that part of Kenya was enough to allow them to rebuild their home, grow pineapples for export, and send their children to school.

The turning point for them, they said, was the opening of a nearby milk-chilling plant. The Talams and other area farmers would take their raw milk to the plant, where it would be kept cold and eventually be transported nationwide, fetching higher prices than it could locally. The plant also served as a kind of training hub. Local dairy farmers could go there to learn how to raise healthier and more productive livestock, get vaccines for their cows, and even have the milk tested for contaminants to make sure it would bring a good price. If it didn't measure up, they got tips on how to improve the quality.

In Kenya, where the Talams live, roughly one-third of the population works in agriculture. Worldwide, there are 500 million smallholder farms, and about two-thirds of people in poverty work in agriculture. Yet despite their large numbers, smallholder farmers are responsible for remarkably few greenhouse gas emissions, because they can't afford to use nearly as many products and services that involve fossil fuels. The typical Kenyan produces 55 times less carbon dioxide than an American, and rural farmers like the Talams produce even less.

But if you remember the problems with cattle that I mentioned in chapter 6, you'll recognize the dilemma right away: The Talams bought more cattle, and cattle contribute more to climate change than any other livestock.

In that respect, the Talams weren't unusual. For many poor farmers, earning more money is a chance to invest in high-value assets, including chickens, goats, and cows—animals that provide good sources of protein and a way to bring in extra cash by selling milk and eggs. It's a sensible decision, and anyone who cares about reducing poverty would hesitate to tell them not to make it. That's the conundrum: As people rise up the income ladder, they do more things that cause emissions. This is why we need innovations—so the poor can improve their lives without making climate change even worse.

The cruel injustice is that even though the world's poor are doing essentially nothing to cause climate change, they're going to suffer the most from it. The climate is changing in ways that will be problematic for relatively well-off farmers in America and Europe, but potentially deadly for low-income ones in Africa and Asia.

As the climate gets warmer, droughts and floods will become more frequent, wiping out harvests more often. Livestock eat less and produce less meat and milk. The air and soil lose moisture, leaving less water available for plants; in South Asia and sub-Saharan Africa, tens of millions of acres of farmland will become substantially drier. Crop-eating pests are already infesting more acreage as they find more hospitable environments to live in. The growing season will also get shorter; at 4 degrees Celsius of warming, most of sub-Saharan Africa could see it shrink by 20 percent or more.

When you're already living on the edge, any one of these changes could be disastrous. If you don't have any money saved up and your crops die off, you can't go buy more seeds; you're just wiped out. What's more, all of these problems will make food far more expensive for those who can least afford it. Because of climate change, prices will skyrocket for hundreds of millions of people who already spend more than half of their incomes on food.

As food becomes less available, an already enormous inequity between rich and poor will get even worse. Today, a child born in Chad is 50 times more likely to die before her fifth birthday than a child born in Finland. With growing food scarcity, more kids won't get all the nutrients they need, weakening their bodies' natural defenses and making them much more likely to die of diarrhea, malaria, or pneumonia. One study found that the number of additional heat-related deaths could approach 10 million a year by the end of the century (that's roughly as many people as are killed by all infectious diseases today), with a large majority of the deaths occurring in poor countries. And the children who don't die will be far more likely to suffer from stunting—that is, to not fully develop physically or mentally.

In the end, the worst impact of climate change in poor countries will be to make health worse—to raise the rates of malnutrition and death. So we need to help the poorest improve their health. I see two ways to do that.

One, we need to raise the odds that malnourished children will survive. That means improving primary health-care systems, doubling down on malaria prevention, and continuing to provide vaccines for conditions like diarrhea and pneumonia. Although the COVID-19 pandemic undoubtedly makes all these things harder, the world knows a lot about how to do them well; the vaccine program known as GAVI, which has prevented 13 million deaths since 2000, ranks as one of humanity's greatest achievements. (The Gates Foundation's contribution to this global undertaking is one of our proudest accomplishments.) We can't let climate change undo this progress. In fact, we need to accelerate it, developing vaccines for other diseases, including HIV, malaria, and tuberculosis, and getting them to everyone who needs them.

Then—in addition to saving the lives of malnourished children—we need to make sure that fewer children are malnourished in the first place. With population growth, the demand for food will likely double or triple in regions where most of the world's poor live. So we need to help poor farmers grow more of it, even when droughts and floods strike. I'll say more about this in the next section.

I spend a lot of time with people who oversee the foreign aid budgets in rich-world countries. Even some very well-intentioned ones have told me, "We used to fund vaccines. Now we need to make our aid budget climate-sensitive"—by which they mean helping Africa lower its greenhouse gas emissions.

I tell them, "Please don't take away vaccine money and put it into electric cars. Africa is responsible for only about 2 percent of all global emissions. What you really should be funding there is *adaptation*. The best way we can help the poor adapt to climate change is to make sure they're healthy enough to survive it. And to thrive despite it."

You've probably never heard of CGIAR.* Neither had I, until a decade or so ago, when I started studying the problems faced by farmers in poor countries. From what I've seen, no other organization has done more than CGIAR to ensure that families—especially the poorest—have nutritious food to eat. And no other organization is in a better position to create the innovations that will help poor farmers adapt to climate change in the years ahead.

CGIAR is the world's largest agricultural research group: In short, it helps create better plants and better animal genetics. It was

---

* CGIAR began life as the Consultative Group for International Agricultural Research. You can see why it started going by the abbreviation.

at a CGIAR lab in Mexico that Norman Borlaug—you may remember him from chapter 6—did his groundbreaking work on wheat, sparking the Green Revolution. Other CGIAR researchers, inspired by Borlaug's example, developed similarly high-yielding, disease-resistant rice, and in the following years the group's work on livestock, potatoes, and maize has helped reduce poverty and improve nutrition.

It's too bad that more people don't know about CGIAR, but it's hardly surprising. For one thing, its name is often mistaken for "cigar," suggesting a link to the tobacco industry. (There isn't one.) And it doesn't help that CGIAR is not a single organization but a network of 15 independent research centers, most of them referred to by their own confusing acronyms. The list includes CIFOR, ICARDA, CIAT, ICRISAT, IFPRI, IITA, ILRI, CIMMYT, CIP, IRRI, IWMI, and ICRAF.

Despite its penchant for alphabet soup, CGIAR will be indispensable in creating new climate-smart crops and livestock for the world's poor farmers. One of my favorite examples is its work on drought-tolerant maize.

Although maize yields in sub-Saharan Africa are lower than anywhere else in the world, more than 200 million households there still depend on this crop for their livelihoods. And as weather patterns have become more erratic, farmers are at greater risk of having smaller maize harvests, and sometimes no harvest at all.

So experts at CGIAR developed dozens of new maize varieties that could withstand drought conditions, each adapted to grow in specific regions of Africa. At first, many smallholder farmers were afraid to try new crop varieties. Understandably so. If you're eking out a living, you won't be eager to take a risk on seeds you've never planted before, because if they die, you have nothing to fall back on. But as experts worked with local farmers and seed dealers to explain the benefits of these new varieties, more and more people adopted them.

The results have been life changing for many families. In Zimbabwe, for example, farmers in drought-stricken areas who used drought-tolerant maize were able to harvest up to 600 more kilograms of maize per hectare than farmers who used conventional varieties. (That's 500 more pounds per acre, producing enough to feed a family of six for nine months.) For farming families who chose to sell their harvests, it was enough extra cash to send their children to school and meet other household needs. CGIAR-affiliated experts have gone on to develop other maize varieties that grow well in poor soils; resist diseases, pests, or weeds; raise crop yields by up to 30 percent; and help fight malnutrition.

And it's not just maize. Thanks to CGIAR's efforts, new types of rice that can tolerate drought are spreading rapidly in India, where climate change is causing more dry spells during the rainy season. They've also developed a type of rice—cleverly nicknamed "scuba" rice—that can survive underwater for two weeks. Generally, rice plants respond to flooding by stretching out their leaves to escape the water; if they're underwater long enough, they expend all their energy trying to escape, and they essentially die of exhaustion. Scuba rice doesn't have that problem: It's got a gene called SUB1 that kicks in during a flood, making the plant dormant—so it stops stretching—until the waters recede.

CGIAR isn't just focused on new seeds. Its scientists have also created a smartphone app that allows farmers to use the camera on their phones to identify specific pests and diseases attacking cassava, an important cash crop in Africa. It's also created programs for using drones and ground sensors to help farmers determine how much water and fertilizer their crops need.

Poor farmers need more advances like these, but to provide them, CGIAR and other agricultural researchers will need more money. Agricultural research is chronically underfunded. In fact, doubling CGIAR's funding so it can reach more farmers is one of the main recommendations by the Global Commission on Adaptation, which

Here's a field planted with scuba rice, which can withstand flooding for two weeks at a time, an advantage that will be even more important as floods become more frequent.

I lead along with the former UN secretary-general Ban Ki-moon and the former World Bank CEO Kristalina Georgieva.* There's no doubt in my mind that this is money well spent: Every dollar invested in CGIAR's research generates about $6 in benefits. Warren Buffett would give his right arm for an investment that paid off six to one, and saved lives in the process.

Aside from helping smallholder farmers raise their crop yields,

---

* The commission is guided by 34 commissioners, including leaders from government, business, nonprofits, and the scientific community; and 19 convening countries, representing all regions of the globe. A global network of research partners and advisers supports the commission. It's co-managed by the Global Center on Adaptation and the World Resources Institute.

our commission on adaptation makes three other recommendations related to agriculture:

Help farmers manage the risks from more chaotic weather. For example, governments can help farmers grow a wider variety of crops and livestock so one setback doesn't wipe them out. Governments should also explore strengthening social-security systems and arranging for weather-based agriculture insurance that helps farmers recover their losses.

Focus on the most vulnerable people. Women aren't the only group of vulnerable people, but they are the biggest. For all sorts of reasons—cultural, political, economic—female farmers have it even harder than men. They may not be able to secure land rights, for example, or have equal access to water, or get financing to buy fertilizer, or even be able to get a weather forecast. So we need to do things like promoting women's property rights and targeting technical advice specifically for them. The payoff could be dramatic: One study by a UN agency found that if women had the same access to resources as men, they could grow 20 to 30 percent more food on their farms and reduce the number of hungry people in the world by 12 to 17 percent.

Factor climate change into policy decisions. Very little money is funneled into helping farmers adapt; only a tiny sliver of the $500 billion that governments spent on agriculture between 2014 and 2016 was directed at activities that will soften the blow of climate change for the poor. Governments should be coming up with policies and incentives to help farmers reduce their emissions while growing more food at the same time.

To sum up: Rich and middle-income people are causing the vast majority of climate change. The poorest people are doing less than anyone else to cause the problem, but they stand to suffer the most from it. They deserve the world's help, and they need more of it than they're getting.

—

The plight of poor farmers—as well as the impact that climate change will have on them—is something I've learned a lot about over the past two decades through my work on global poverty. It's also a passion of mine, because I get to geek out on the fascinating science behind plant breeding.

Until recently, though, I hadn't put as much thought into other pieces of the adaptation puzzle, like what cities should do to prepare or how ecosystems will be affected. But lately I've had the chance to go deeper through my work with the commission on adaptation I just mentioned. Here are a few insights I've gained from the commission's work—informed by dozens of experts in science, public policy, industry, and other areas—to give you a sense of what else it's going to take to adapt to a warmer climate.

Broadly speaking, you can think of adaptation in three stages. The first involves reducing the risks posed by climate change, through steps like climate-proofing buildings and other infrastructure, protecting wetlands as a bulwark against flooding, and—when necessary—encouraging people to relocate permanently from areas that are no longer livable.

Next is preparing for and responding to emergencies. We need to keep improving weather forecasts and early-warning systems for getting out information about storms. And when disaster does strike, we need well-equipped and well-trained teams of first responders and a system in place for handling temporary evacuations.

Finally, after a disaster, there's the recovery period. We'll need to plan for services for people who've been displaced—services like health care and education—as well as insurance that helps people at all income levels rebuild and standards to ensure that whatever gets rebuilt is more climate-proof than what was there before.

Here are four of the big headlines on adaptation:

Cities need to change the way they grow. Urban areas are home to more than half the people on earth—a proportion that will rise in the years ahead—and they're responsible for more than three-quarters of the world's economy. As they expand, many of the world's fast-growing cities end up building over floodplains, forests, and wetlands that could absorb rising waters during a storm or hold reservoirs of water during a drought.

All cities will be affected by climate change, but coastal cities will have the worst problems. Hundreds of millions of people could be forced from their homes as sea levels rise and storm surges get worse. By the middle of this century, the cost of climate change to all coastal cities could exceed $1 trillion . . . each year. To say that this will exacerbate the problems most cities are already struggling with—poverty, homelessness, health care, education—would be an understatement.

What does climate-proofing a city look like? For one thing, city planners need the latest data on climate risks and projections from computer models that predict the impact of climate change. (Today, many city leaders in the developing world don't have even basic maps to indicate which areas of town are most prone to floods.) Armed with the latest information, they can make better decisions about how to plan for neighborhoods and industrial centers, build or expand seawalls, protect themselves from the storms that are getting more violent, shore up storm-water drainage systems, and raise wharves so they stay above rising tides.

To get really specific: If you're building a bridge across the local river, should you make it 12 feet tall or 18 feet tall? The taller one will be more expensive in the short run, but if you know the odds are high that a massive flood will come along in the next decade, it could be the smarter choice. You'd rather build a more expensive bridge once than a cheaper bridge twice.

And it's not just about renovating the infrastructure that cities already have; climate change is also going to force us to consider entirely new needs. For example, cities with extremely hot days and a lot of residents who can't afford air-conditioning will need to create cooling centers—facilities where people can go to escape the heat. Unfortunately, using more air conditioners also means we'll be emitting more greenhouse gases, which is another reason why the advances in cooling that I discussed in chapter 8 are so important.

We should shore up our natural defenses. Forests store and regulate water. Wetlands prevent floods and provide water for farmers and cities. Coral reefs are home to the fish that coastal communities depend on for food. But these and other natural defenses against climate change are rapidly disappearing. Nearly nine million acres of old-growth forest were destroyed in 2018 alone, and when—as is likely—we hit 2 degrees Celsius of warming, most of the coral reefs in the world will die off.

On the other hand, restoring ecosystems has a huge payoff. Water utilities in the world's largest cities could save $890 million a year by restoring forests and watersheds. Many countries are already leading the way: In Niger, one local reforestation effort led by farmers has boosted crop yields, increased tree cover, and cut the amount of time women spend gathering firewood from three hours a day to 30 minutes. China has identified about a quarter of its landmass as critical natural assets where it'll make a priority of promoting conservation and preserving the ecosystem. Mexico is protecting a third of its river basins to preserve the water supply for 45 million people.

If we can build on these examples, spreading awareness about how much ecosystems matter and helping more countries follow suit, we'll gain the benefits of a natural defense against climate change.

Here's some more low-hanging fruit, so to speak: mangrove forests. Mangroves are short trees that grow along coastlines, having adapted to life in salt water; they reduce storm surges, prevent coastal flooding, and protect fish habitats. All told, mangroves help the world avoid some $80 billion a year in losses from floods, and

Planting mangrove trees is a great investment. They help prevent some $80 billion a year in losses from floods.

they save billions more in other ways. Planting mangroves is much cheaper than building breakwaters, and the trees also improve the water quality. They're a great investment.

We're going to need more drinking water than we can supply. As lakes and aquifers shrink or get polluted, it's getting harder to provide potable water to everyone who needs it. Most of the world's megacities already face severe shortages, and if nothing changes, by mid-century the number of people who can't get enough decent water at least once a month will rise by more than a third, to over 5 billion people.

Technology holds out some promise here. We already know how to take the salt out of seawater and make it drinkable, but the process takes a lot of energy, as does moving the water from the ocean to the desalination facility and then from the facility to whoever needs it. (This means that, like so many things, the water problem is ultimately an energy problem: With enough cheap, clean energy, we can make all the potable water we'll ever need.)

One clever idea I'm watching closely involves taking water out of the air. It's basically a solar-powered dehumidifier with an advanced filtering system so you don't drink air pollution. This system is available now, but it costs thousands of dollars, far too expensive for the world's poor, who will suffer the most from water shortages.

Until an idea like that becomes affordable, we need to take practical steps—incentives that will drive the demand for water down and efforts that will drive the supply up. That includes everything from reclaiming wastewater to just-in-time irrigation, a system that reduces water use dramatically while raising farmers' yields.

Finally, to fund adaptation projects, we need to unlock new money. I'm talking not about foreign aid for developing countries—although we'll need that too—but about how public money can attract private investors to get behind adaptation projects.

Here's the problem we need to overcome: People pay the costs of adaptation up front, but its economic benefits may not come for years down the road. For example, you can flood-proof your business now, but it may not get hit by a big deluge for 10 or 20 years. And your flood-proofing isn't going to generate bankable cash flows; customers aren't going to pay extra for your products because you made sure sewage won't back up into your basement during a flood. So banks will be reluctant to loan you the money for your project, or they'll charge you a higher interest rate. Either way, you have to absorb some cost yourself, in which case you may simply decide not to do it.

Take that single example and multiply it across an entire city, state, or country, and you'll see why the public has to play a role in both financing adaptation projects and drawing in the private sector as well. We need to make adaptation an attractive investment.

That starts with finding ways for public and private financial markets to take the risks of climate change into account and to price these risks accordingly. Some governments and companies already screen their projects for climate risks; all of them should. Governments can also put more resources into adaptation, setting goals for how much they'll invest over time and adopting policies that remove some of the risk for private investors. As the rewards of adaptation projects become more clear, private investment should grow.

You may be wondering how much all this would cost. There's no way to put a price tag on everything the world needs to do to adapt to climate change. But the commission I'm involved with priced out spending in five key areas (creating early-warning systems, building climate-resilient infrastructure, raising crop yields, managing water, and protecting mangroves) and found that investing $1.8 trillion between 2020 and 2030 would return more than $7 trillion in benefits. To put that in perspective, spread out over a decade, it's about 0.2 percent of the world's GDP, with a nearly fourfold return on investment.

You can measure those benefits in terms of bad things that don't happen: civil wars that don't break out over water rights, farmers who don't get wiped out by a drought or flood, cities that don't get destroyed by hurricanes, waves of people who don't become climate refugees. Or you can measure them in terms of good things that *do* happen: children who grow up with the nutrients they need, families who escape poverty and join the global middle class, businesses and cities and countries that thrive even as the climate gets hotter.

Whichever way you think about it, the economic case is clear, and so is the moral case. Extreme poverty has plummeted in the past quarter century, from 36 percent of the world's population in 1990 to 10 percent in 2015—although COVID-19 was a huge setback that undid a great deal of progress. Climate change could erase even more of these gains, increasing the number of people living in extreme poverty by 13 percent.

Those of us who have done the most to cause this problem should help the rest of the world survive it. We owe them that much.

—

There's one other aspect to adaptation that deserves a lot more attention than it's getting: We need to be preparing for a worst-case scenario.

Climate scientists have identified many tipping points that could dramatically increase the rate at which climate change happens—for instance, if the ice-like crystalline structures containing large amounts of methane on the ocean floor become unstable and erupt. In a relatively short time, disasters could strike around the world, overwhelming our attempts to prepare for and respond to climate change. And the higher the temperature goes, the more likely we are to reach a tipping point.

If it starts looking as if we're headed toward one of these tipping points, you're going to hear more about a set of bold—some would say crazy—ideas that fall under the umbrella term "geoengineering." These approaches are unproven, and they raise thorny ethical issues. But they're worth studying and debating while we still have the luxury of study and debate.

Geoengineering is a cutting-edge, "Break Glass in Case of Emergency" kind of tool. The basic idea is to make temporary changes in the earth's oceans or atmosphere that lower the planet's temperature. These changes wouldn't be intended to absolve us of the responsibility to reduce emissions; they'd just buy us time to get our act together.

For a few years, I've been funding some studies on geoengineering (this funding is tiny compared with the work on mitigation and adaptation that I'm supporting). Most approaches to geoengineering are based on the idea that to compensate for all the warming caused by greenhouse gases we've added to the atmosphere, we need to reduce the amount of sunlight hitting the earth by around 1 percent.*

---

* If you want to know the math: Sunlight is absorbed by the earth at a rate of about 240 watts per square meter. There's enough carbon in the atmosphere now

There are various ways we could do that. One involves distributing extremely fine particles—each just a few millionths of an inch in diameter—in the upper layers of the atmosphere. Scientists know that these particles would scatter sunlight and cause cooling, because they've watched it happen: When an especially powerful volcano erupts, it spews out a similar type of particle and measurably drives down the global temperature.

Another approach to geoengineering involves brightening clouds. Because sunlight is scattered by the tops of clouds, we could scatter more sunlight and cool the earth by making the clouds brighter, using a salt spray that causes clouds to scatter more light. And it wouldn't take a dramatic increase; to get the 1 percent reduction, we'd only need to brighten clouds that cover 10 percent of the earth's area by 10 percent.

There are other approaches to geoengineering; they all have three things in common. One, they're relatively cheap compared with the scale of the problem, requiring up-front capital costs of less than $10 billion and minimal operating expenses. Two, the effect on clouds lasts for a week or so, so we could use them as long as we needed to and then stop with no long-term impacts. And three, whatever technical problems these ideas might face are nothing compared with the political hurdles they'll definitely face.

Some critics attack geoengineering as a massive experiment on the planet, though as the proponents of geoengineering point out, we're already running a massive experiment on the planet by emitting huge amounts of greenhouse gases.

---

to absorb heat at an average rate of about 2 watts per square meter. So we need to make the sun dimmer by 2/240, or 0.83 percent. However, because clouds would adjust to solar geoengineering, we would actually need to dim the sun a bit more, to about 1 percent of the incoming sunlight. If the amount of carbon in the atmosphere doubles, it would absorb heat at a rate of about 4 watts per square meter, and we would need to double the dimming to about 2 percent.

What's fair to say is that we need to better understand the potential impact of geoengineering at a local level. That's a legitimate concern that deserves much more study before we even consider testing geoengineering at scale in the real world. Also, because the atmosphere is literally a global concern, no single nation could decide to try geoengineering on its own. We'd need some consensus.

Right now, it's hard to imagine getting countries around the world to agree to artificially set the planet's temperature. But geoengineering is the only known way that we could hope to lower the earth's temperature within years or even decades without crippling the economy. There may come a day when we don't have a choice. Best to prepare for that day now.

CHAPTER 10

WHY GOVERNMENT POLICIES MATTER

n 1943, at the height of World War II, a thick cloud of smoke descended on Los Angeles. It was so noxious that it made residents' eyes sting and their noses run. Drivers couldn't see more than three blocks down the road. Some locals feared that the Japanese army had attacked the city with chemical weapons.

L.A. hadn't been attacked, though—at least, not by a foreign army. The real culprit was smog, created by an unfortunate combination of air pollution and weather conditions.

Almost a decade later, for five days in December 1952, London too was crippled by smog. Buses and ambulances stopped running. Visibility was so low, even within enclosed buildings, that movie theaters were shut down. Looting was rampant because the police couldn't see more than a few feet in any direction. (If you're a fan of the Netflix series *The Crown*, as I am, you'll remember a gripping episode in season 1 that takes place during this awful incident.) What's now known as the Great Smog of London killed at least 4,000 people.

Thanks to incidents like these, the 1950s and 1960s marked the arrival of air pollution as a major cause of public concern in the United States and Europe, and policy makers responded quickly. Congress began to provide funding for research into the problem and possible remedies in 1955. The next year, the British government

This police officer had to use a flare to direct traffic during the Great Smog of London in 1952.

enacted the Clean Air Act, which created smoke-control zones throughout the country where only cleaner-burning fuels could be used. Seven years later, America's Clean Air Act established the modern regulatory system for controlling air pollution in the United States; it remains the most comprehensive law—and one of the most influential—to regulate air pollution that can endanger public health. In 1970, President Nixon established the Environmental Protection Agency to help implement it.

The U.S. Clean Air Act did what it was supposed to do—get poisonous gases out of the air—and since 1990 the level of nitrogen dioxide in American emissions has dropped by 56 percent, carbon monoxide by 77 percent, and sulfur dioxide by 88 percent. Lead has nearly vanished from American emissions. While we still have work to do, we accomplished all this even as our economy and population grew.

But you don't have to look to history for examples of how smart policies help solve a problem like air pollution. It's happening right now. Starting in 2014, China launched several programs in response to worsening smog in urban centers and skyrocketing levels of dangerous air pollutants. The government set new targets for reducing air pollution, banned the building of new coal-fired plants near the most polluted cities, and put limits on driving nonelectric cars in large cities. Within a few years, Beijing was reporting a 35 percent decline in certain types of pollution, and Baoding, a city of 11 million people, was reporting a decline of 38 percent.

Although air pollution is still a major cause of illness and death—it likely kills more than 7 million people every year—the policies we've put in place have undoubtedly kept the number from being even higher.* (They've also helped reduce greenhouse gases a bit, even though that wasn't their original purpose.) Today they illustrate as well as anything the leading role that government policies have to play in avoiding a climate disaster.

I admit that "policy" is a vague, dull-sounding word. A big breakthrough like a new type of battery would be sexier than the policies that led some chemist to invent it. But the breakthrough wouldn't even exist without a government spending tax dollars on research, policies designed to drive that research out of the lab and into the market, and regulations that created markets and made it easy to deploy at scale.

In this book, I've been emphasizing the inventions we need to get to zero—new ways of storing electricity, making steel, and so on—but innovation is not just a matter of developing new devices. It's also a matter of developing new policies so we can demonstrate and deploy those inventions in the market as fast as possible.

Luckily, in developing these policies, we're not starting with a blank slate. We've got a *lot* of experience regulating energy. In fact,

---

* Wildfires, such as the ones that swept across the western United States in 2020, are a separate but related issue. Smoke from the 2020 wildfires made it unsafe for millions of people to go outside.

it's one of the most heavily regulated sectors of the economy, in the United States and around the world. In addition to cleaner air, smart energy policies have given us the following:

Electrification. In 1910, only 12 percent of Americans had electric power in their homes. By 1950, more than 90 percent did, thanks to efforts like federal funding for dams, the creation of federal agencies to regulate energy, and a massive government project to bring electricity to rural areas.

Energy security. In response to the oil shocks of the 1970s, the United States set out to increase domestic production from various energy sources. The federal government began its first major research and development projects in 1974. The next year saw major legislation related to energy conservation, including fuel efficiency standards for cars. Two years later came the creation of the Department of Energy. Then, in the 1980s, oil prices collapsed, and we abandoned many of these efforts—until prices started rising again in the 2000s, sparking a new wave of investment and regulation. As a result of these and other efforts, in 2019 America exported more energy than it imported for the first time in nearly 70 years.

Economic recovery. After the Great Recession of 2008, governments created jobs and spurred investment by putting money into renewable energy, energy efficiency, electricity infrastructure, and railroads. In 2008, China launched a $584 billion economic stimulus package, a large part of which went to green projects. In 2009, the American Recovery and Reinvestment Act used tax credits, federal grants, loan guarantees, and R&D funding to shore up the economy and reduce emissions. This was the single largest investment in clean energy and energy efficiency in American history, but it was a onetime injection, not a lasting change in policy.

Now it's time to turn our policy-making experience to the challenge at hand: zeroing out our greenhouse gas emissions.

National leaders around the world will need to articulate a vision for how the global economy will make the transition to zero carbon. That vision can, in turn, guide the actions of people and businesses around the world. Government officials can write rules regarding how much carbon power plants, cars, and factories are allowed to emit. They can adopt regulations that shape financial markets and clarify the risks of climate change to the private and public sectors. They can be the main investors in scientific research, as they are now, and write the rules that determine how quickly new products can get to market. And they can help fix some problems that the market isn't set up to deal with—including the hidden costs that carbon-emitting products impose on the environment and on humans.

Many of these decisions are made at the national level, but state and local governments have a big role too. In many countries, subnational governments regulate electricity markets and set standards for energy use in buildings. They plan massive construction projects—dams, transit systems, bridges, and roads—and choose where these projects will be built and with what materials. They buy police cars and fire engines, school lunches, and lightbulbs. At each step, someone will have to decide whether to go with the green alternatives.

It might seem ironic that I'm calling for more government intervention. When I was building Microsoft, I kept my distance from policy makers in Washington, D.C., and around the world, thinking they would only keep us from doing our best work.

In part, the U.S. government's antitrust suit against Microsoft in the late 1990s made me realize that we should've been engaging with policy makers all along. I also know that when it comes to massive undertakings—whether it's building a national highway system, vaccinating the world's children, or decarbonizing the global economy—we need the government to play a huge role in creating the right incentives and making sure the overall system will work for everyone.

Of course, businesses and individuals will need to do their part too. In chapters 11 and 12, I'll propose a plan for getting to zero, with specific steps that governments, businesses, and individuals can take. But because governments will play such a major role, first I want to suggest seven high-level goals they should be aiming for.

1. Mind the Investment Gap

The first microwave oven hit the market in 1955. It cost, in today's dollars, nearly $12,000. Today, you can get a perfectly good one for $50.

Why did microwaves get so cheap? Because it was immediately obvious to consumers why you'd want something that could heat up food in a fraction of the time it took your conventional oven. Microwave sales rose quickly, which drove competition in the marketplace, which led to the production of cheaper and cheaper appliances.

If only the energy market worked the same way. Electricity isn't like a microwave oven, where the product with the best features wins out. A dirty electron will run your lights just as well as a clean one. As a result, without some policy intervention—such as a price on carbon, or standards that require a certain volume of zero-carbon electrons in the marketplace—there's no guarantee that the company that invests in sending you clean electrons will actually make money. And there's considerable risk, because energy is such a highly regulated and capital-intensive industry.

So you can see why the private sector systematically underinvests in R&D on energy. Companies in the energy business spend an average of just 0.3 percent of their revenue on energy R&D. The electronics and pharmaceutical industries, by contrast, spend nearly 10 percent and 13 percent, respectively.

We'll need government policies and financing to close the gap, focusing especially in areas where we need to invent new zero-carbon technologies. When an idea is in its earliest stages—when we're not sure whether it'll work, and success may take longer than banks or venture capitalists are willing to wait—the right policies and financing can make sure it gets fully explored. It might be a big breakthrough, but it might be a bust, so we'll need to tolerate some outright failures.

In general, the government's role is to invest in R&D when the private sector won't because it can't see how it will make a profit. Once it becomes clear how a company can make money, the private sector takes over. This is in fact exactly how we got products you probably use every day, including the internet, lifesaving medicines, and the Global Positioning System that your smartphone uses to help you navigate around town. The personal computer business—including Microsoft—would never have been the success that it was if the U.S. government hadn't put money into research on smaller, faster microprocessors.

In some sectors, like digital technology, the government-to-company handoff happens relatively quickly. With clean energy, it takes much longer and requires even more financial commitment from the government, because the scientific and engineering work is so time-consuming and expensive.

Investing in research has another benefit: It can help create businesses in one country that export their products to others. Country 1, for example, could create a cheap electrofuel, selling it to its own people but also exporting it to country 2. Even if country 2 otherwise lacks the ambition to reduce its emissions, it will end up doing so, simply because someone else invented a better, cheaper fuel.

Finally, although R&D yields benefits on its own, it is most effective when you pair it with demand-side incentives. No business is going to turn that new idea published in a scientific journal into a product unless it's confident that it'll find willing buyers, particularly in the early stages, when the product will be expensive.

2. Level the Playing Field

As I've argued ad infinitum (and possibly ad nauseam), we need to reduce the Green Premiums to zero. Some of that we can accomplish with the innovations I described in chapters 4 through 8—by making it cheaper to produce zero-carbon steel, for instance. But we can also raise the cost of fossil fuels by incorporating the damage they cause into the prices we pay for them.

Today, when businesses make products or consumers buy things, they don't bear any extra cost for the carbon involved, even though that carbon imposes a very real cost on society. This is what economists call an externality: an expense that's borne by society rather than the person or business who's responsible for it. There are various ways, including a carbon tax or cap-and-trade program, to ensure that at least some of these external costs are paid by whoever is responsible for them.

In short, we can reduce Green Premiums by making carbon-free things cheaper (which involves technical innovation), by making carbon-emitting things more expensive (which involves policy innovation), or by doing some of both. The idea isn't to punish people for their greenhouse gases; it's to create an incentive for inventors to create competitive carbon-free alternatives. By progressively increasing the price of carbon to reflect its true cost, governments can nudge producers and consumers toward more efficient decisions and encourage innovation that reduces Green Premiums. You're a lot more likely to try to invent a new kind of electrofuel if you know it won't be undercut by artificially cheap gasoline.

3. Overcome Nonmarket Barriers

Why are homeowners reluctant to abandon fossil-fuel-powered furnaces in favor of lower-emissions electric options? Because they don't know about the alternatives, there aren't enough qualified dealers and installers to provide them, and in some places it's actually illegal.

Why don't landlords upgrade their buildings with more efficient appliances? Because they pass the energy bills on to their tenants, who often aren't allowed to make the upgrades and who probably won't live there long enough to reap the long-term benefits anyway.

Neither of these barriers, you'll notice, has much to do with cost. They exist mainly because of a lack of information, or trained personnel, or incentives—all areas in which the right government policies can make a big difference.

4. Stay Up to Date

Sometimes the big barrier isn't consumer awareness or markets that are out of whack. Sometimes it's government policies themselves that make it hard to decarbonize.

For example, if you want to use concrete in a building, the building code will spell out in excruciating detail how well that concrete has to perform—how strong it has to be, how much weight it can bear, and so on. It may also define the precise chemical composition of the concrete you can use. These composition standards often rule out a low-emissions cement that you want to use, even if it meets all the performance standards.

No one wants to see buildings and bridges collapsing because of faulty concrete. But we can make sure the standards reflect the latest advances in technology and the urgency of getting to zero.

5. Plan for a Just Transition

Such a massive shift to a carbon-neutral economy is bound to produce winners and losers. In the United States, states whose economies rely heavily on drilling for fossil fuels—Texas and North Dakota, for example—will need to add jobs that pay as well as the ones they lose, and they'll need to replace tax revenue that currently pays for schools, roads, and other essentials. So will beef-growing states like Nebraska, if artificial meats take the place of conventional ones. And low-income people, who already spend a significant portion of their income on energy, will feel the burden of Green Premiums more than most.

I wish there were easy answers to these problems. Certainly there are some communities where high-paying oil and gas jobs will naturally be replaced with jobs in, say, the solar industry. But many others will need to go through a difficult transition to relying on something other than extracting fossil fuels for their livelihood. Because the solutions will vary from place to place, they'll need to be shaped by local leaders. But the federal government can help—as part of an overall plan for getting to zero—by providing funding and technical advice and by connecting communities around the country that are experiencing similar problems so they can share what's working.

Finally, in communities where extracting coal or natural gas is a big part of the local economy, it's understandable that people worry about how the transition might make it harder for them to make ends meet. The fact that they voice those worries doesn't make them climate change deniers. You don't have to be a political scientist to think that national leaders who champion getting to zero will find more support for their ideas if they understand the concerns of families and communities whose livelihoods will be hit hard and if they take those concerns seriously.

6. Do the Hard Stuff Too

A lot of climate change work focuses on the relatively easy ways to reduce emissions—things like driving electric cars and getting more power from solar and wind. That makes sense, because showing progress and demonstrating early success helps get more people on board. And it's important: We're not doing the relatively easy stuff at nearly the scale we need, so there are huge opportunities to make major progress right now.

But we can't just go after this low-hanging fruit. Now that the movement to address climate change is getting serious, we'll need to focus on the hard parts too: electricity storage, clean fuels, cement, steel, fertilizer, and so on. And that will require a different approach to policy making. In addition to deploying the tools we already have, we'll need to invest more in R&D on the hard stuff and—because much of it is core to our physical infrastructure, like roads and buildings—use policies specifically designed to get these breakthroughs created and into the marketplace.

7. Work on Technology, Policy, and Markets at the Same Time

In addition to technology and policy, there's a third aspect that we'll need to factor in: the companies that will develop new inventions and make sure they reach a global scale, as well as the investors and financial markets that will back these companies. For lack of a better term, I'll broadly call this group "markets."

Markets, technology, and policy are like three levers that we need to pull in order to wean ourselves from fossil fuels. We need to pull all three of them at the same time and in the same direction.

Simply adopting a policy—say, a zero-emissions standard for cars—won't do much good if you don't have the technology to eliminate emissions or if there aren't any companies willing to manufacture and sell cars that meet the standard. On the other hand, having a low-emissions technology—say, a device that captures carbon from a coal plant's exhaust—won't do much good if you don't create the financial incentive for power companies to install it. And few companies will make a bet on inventing zero-emissions technology if their competitors can undersell them with fossil-fuel products.

That's why markets, policy, and technology have to work in complementary ways. Policies, such as higher spending on R&D, can help spark new technologies and shape the market systems that will make sure they reach millions of people. But it works the other way too: Policies should also be shaped by the technologies we develop. If, for example, we came up with a breakthrough liquid fuel, then our policies would focus on creating the investment and financing strategies to get it to global scale, and we wouldn't need to worry as much about, say, finding new ways to store energy.

I'll give you a few examples of what happens when all three things work together and when they don't.

To see the effect of policies that don't keep up with technology, look at the nuclear power industry. Nuclear is the only carbon-free energy source we can use almost anywhere, 24 hours a day, 7 days a week. A handful of companies, including TerraPower, are working on advanced reactors that solve the problems of the 50-year-old design used by reactors you see today: Their designs are safer and cheaper and produce much less waste. But without the right policies and the right approach to markets, the scientific and engineering work on these advanced reactors will go nowhere.

No advanced nuclear plant will ever be built unless the design can be validated, the supply chains can be established, and a pilot project can be built to demonstrate the new approach. Unfortunately, with a few exceptions like China and Russia—which are directly investing in state-supported advanced nuclear companies—most countries don't have workable ways to do these things. It would help if some governments were willing to co-invest in them to help get demonstration projects up and running—as the U.S. government has done recently. I realize this might sound self-serving, given that I own an advanced-nuclear company, but it's the only way nuclear power will have a chance of helping with climate change.

The example of biofuels shows a different challenge: making sure we know what problem we're trying to solve, and tuning our policies accordingly.

In 2005, with an eye on rising oil prices and a desire to import less oil, Congress passed the Renewable Fuel Standard, which set targets for how much biofuel the country would use in the coming years. Simply passing this legislation sent a strong signal to the transportation industry, which invested a lot in the biofuel technology that existed at the time—corn-based ethanol. Corn ethanol was already fairly competitive with gasoline, because gas prices were going up and ethanol producers benefited from a decades-old tax credit.

The policy worked. Ethanol production quickly exceeded the targets that Congress had set; today a gallon of gas sold in the United States might contain up to 10 percent ethanol.

Then, in 2007, Congress set out to use biofuels to solve a different problem. Now the concern wasn't just rising oil prices; it was climate change too. The government raised the biofuels targets and in addition required that about 60 percent of all the biofuels sold in the United States be made not from corn but from other starches. (Biofuels made in this way reduce emissions three times more than conventional biofuels.) Refiners quickly met the target for conventional corn-based biofuels, but the advanced alternatives have lagged far behind their target.

Why? Partly because the science of advanced biofuels is just plain hard. And oil prices have stayed relatively low, making it difficult to justify major investments in an alternative that will be more expensive. But a big reason is that the companies that might produce these biofuels, and the investors who might back them, haven't had any certainty about the market.

The executive branch has expected shortfalls in the supply of advanced biofuels, so it keeps lowering the targets. In 2017, the target was dropped from 5.5 billion gallons to 311 million gallons. And sometimes the new targets are announced so late in the year that producers don't know how much they can count on selling. It's a vicious cycle: The government lowers the quota because it expects a shortfall, and the shortfalls keep happening because the government keeps lowering the quota.

The lesson here is that policy makers need to be clear about the goal they're trying to achieve and aware of the technologies they're trying to promote. Setting a biofuels target was a fine way to reduce the amount of oil the United States needed to import, because there was already an existing technology—corn ethanol—that could meet the target. The policy sparked innovation, developed the market, and got it to scale up. But setting a biofuels target wasn't a particularly effective way to lower emissions, because policy makers haven't accounted for the fact that the suitable technology—advanced biofuels—is still in the early stages and they haven't created the certainty that the market needs to get it out of the early stages.

Now let's look at a success story where policy, technology, and markets worked together much better. As early as the 1970s, Japan, the United States, and the European Union began funding early research into various ways of generating electricity from sunlight. By the early 1990s, solar technology had improved enough that more companies started making panels, but solar still wasn't being widely adopted.

Germany gave the market a boost by offering low-interest loans to install panels and paying a feed-in tariff—a fixed government payment per unit of electricity generated by renewables—to anyone who generated excess solar power. Then, in 2011, the United States used loan guarantees to finance the five biggest solar arrays in the country. China has been a major player in finding ingenious ways to make solar panels cheaper. Thanks to all this innovation, the price of solar-generated electricity has dropped 90 percent since 2009.

Wind power is another good example. Over the past decade, installed wind capacity has grown by an average of 20 percent a year, and wind turbines now provide about 5 percent of the world's electricity. Wind is growing for one simple reason: It's getting cheaper. China, which accounts for a large and growing share of the world's wind-generated power, has said it will soon stop subsidizing onshore wind projects because the electricity they produce will be just as cheap as the power from conventional sources.

To understand how we got to this point, look at Denmark. Amid the oil shocks of the 1970s, the Danish government enacted a number of policies with an eye toward promoting wind energy and importing less oil. Among other things, the government put a lot of money into renewable-energy R&D. They weren't the only ones who did this (around this time, the United States started working on

Denmark helped lead the way on making wind power more affordable. These turbines are on the island of Samsø.

utility-scale wind turbines in Ohio), but the Danes did something unusual. They paired their R&D support with a feed-in tariff and, later, a carbon tax.

As countries like Spain followed suit, the wind industry started moving down the learning curve. Companies now had the incentive to develop larger rotors and higher-capacity machines so each turbine could produce more power, and they started selling more units. Over time, the cost of a wind turbine dropped dramatically. So did the cost of the electricity generated by wind: In Denmark, it fell by half between 1987 and 2001. Today, the country gets about half its electricity from onshore and offshore wind, and it's the largest exporter of wind turbines in the world.

To be clear: The point of these stories is not that solar and wind are the answer to all our electricity needs. (They are two of the answers to some of our electricity needs. See chapter 4.) The point is that when we focus on all three things at once—technology, policies, and markets—we can encourage innovation, spark new companies, and get new products into the market fast.

Any plan for climate change needs to understand how all three work together. In the next chapter, I'll propose one that does just that.

CHAPTER 11

A PLAN FOR GETTING TO ZERO

When I was in Paris in 2015 for the climate conference, I couldn't help wondering: Can we really do this?

It was inspiring to see leaders from around the world come together to embrace climate goals as nearly every nation committed to cut its emissions. But with one poll after another showing that climate change was still a marginal political issue (at best), I worried that we'd never have the will to do this hard job.

I'm glad to say that the public's interest in climate change has grown much more than I thought it would. Over the past few years, the global conversation about climate change has taken a remarkable turn for the better. Political will is growing at every level as voters around the world demand action and cities and states commit to making dramatic reductions that support (or, as in the United States, fill in for) their national goals.

Now we need to pair these goals with specific plans for achieving them—just as, in the early days of Microsoft, Paul Allen and I had a goal ("a computer on every desk and in every home," as we put it) and spent the next decade building and executing a plan to reach it. People thought we were crazy to dream so big, but that challenge was nothing compared with what it'll take to deal with climate change, a massive undertaking that will involve people and institutions around the world.

Chapter 10 was all about the role governments need to play in achieving that goal. In this chapter, I'll propose a plan for how we can avoid a climate disaster, focusing on the specific steps government leaders and policy makers can take. (You can find more detail on each element below at breakthroughenergy.org.) In the next chapter, I'll lay out what each of us can do as individuals to support this plan.

How quickly do we need to get to zero? Science tells us that in order to avoid a climate catastrophe, rich countries should reach net-zero emissions by 2050. You've probably heard people say we can decarbonize deeply even sooner—by 2030.

Unfortunately, for all the reasons I've laid out in this book, 2030 is not realistic. Considering how fundamental fossil fuels are in our lives, there's simply no way we'll stop using them widely within a decade.

What we can do—and *need* to do—in the next 10 years is adopt the policies that will put us on a path to deep decarbonization by 2050.

This is a crucial distinction, though it's not one that's immediately obvious. In fact, it might seem like "reduce by 2030" and "get to zero by 2050" are complementary. Isn't 2030 a stop on the way to 2050?

Not necessarily. Making reductions by 2030 the wrong way might actually *prevent* us from ever getting to zero.

Why? Because the things we'd do to get small reductions by 2030 are radically different from the things we'd do to get to zero by 2050. They're really two different pathways, with different measures of success, and we have to choose between them. It's great to have goals for 2030, as long as they're milestones on the way to zero by 2050.

Here's why. If we set out to reduce emissions only somewhat by 2030, we'll focus on the efforts that will get us to that goal—even if those efforts make it harder, or impossible, to reach the ultimate goal of getting to zero.

For example, if "reduce by 2030" is the only measure of success, then it would be tempting to replace coal-fired power plants with gas-fired ones; after all, that would reduce our emissions of carbon dioxide. But any gas plants built between now and 2030 will still be in operation come 2050—they have to run for decades in order to recoup the cost of building them—and natural gas plants still produce greenhouse gases. We would meet our "reduce by 2030" goal but have little hope of ever getting to zero.

On the other hand, if our "reduce by 2030" goal is a milestone toward "zero by 2050," then it makes little sense to spend a lot of time or money switching from coal to gas. Instead, we're better off pursuing two strategies at the same time: First, going all out to deliver zero-carbon electricity cheaply and reliably; and second, electrifying as widely as possible—everything from vehicles to industrial processes and heat pumps, even in places that currently rely on fossil fuels for their electricity.

If we think the only thing that matters is reducing emissions by 2030, then this approach would be a failure, since it might deliver only marginal reductions within a decade. But we'd be setting ourselves up for long-term success. With every breakthrough in generating, storing, and delivering clean electricity, we would march closer and closer to zero.

So if you want a measuring stick for which countries are making progress on climate change and which ones aren't, don't simply look for the ones that are reducing their emissions. Look for the ones that are setting themselves up to get to zero. Their emissions might not be changing much now, but they deserve credit for getting on the right path.

I agree with the 2030 advocates on one thing: This is urgent work. We are at the same point today with climate change as we were several years ago with pandemics. Health experts were telling us that a massive outbreak was virtually inevitable. Despite their warnings, the world didn't do enough to prepare—and then suddenly had to scramble to make up for lost time. We should not make the same mistake with climate change. Given that we'll need these breakthroughs before 2050, and given what we know about how long it takes to develop and roll out new energy sources, we need to start now. If we do start now, tapping into the power of science and innovation and ensuring that solutions work for the poorest, we can avoid repeating the mistakes of pandemic preparation with climate change. This plan sets us on that path.

Innovation and the Law of Supply and Demand

As I argued at the outset—and as I hope has become clear in the intervening chapters—any comprehensive climate plan has to tap into many different disciplines. Climate science tells us *why* we need to deal with this problem but not *how* to deal with it. For that, we'll need biology, chemistry, physics, political science, economics, engineering, and other sciences. Not that everyone needs to understand every subject, any more than Paul and I were experts at marketing, partnering with businesses, or working with governments when we started out. What Microsoft needed—and what we need now to deal with climate change—is an approach that allows many different disciplines to put us on the right path.

In energy, software, and just about every other pursuit, it's a mistake to think of innovation only in the strict, technological sense. Innovation is not just a matter of inventing a new machine or some new process; it's also coming up with new approaches to business models, supply chains, markets, and policies that will help new inventions come to life and reach a global scale. Innovation is both new devices and new ways of doing things.

With those provisos in mind, I've divided the different elements of my plan into two categories. These categories will sound familiar if you've taken Economics 101: One involves expanding the *supply* of innovations—the number of new ideas that get tested—and the other involves accelerating the *demand* for innovations. The two work hand in hand, in a push-and-pull fashion. Without demand for innovation, inventors and policy makers won't have any incentive to push out new ideas; without a steady supply of innovations, buyers won't have the green products the world needs to get to zero.

I realize this may sound like business-school theory, but it's actually quite practical. The Gates Foundation's whole approach to saving lives is based on the idea that we need to be pushing innovation for the poor while also increasing demand for it. And at Microsoft, we created a large group that did nothing but research, something I'm proud of to this day. Essentially, their job is to increase the supply of innovations. We also spent a great deal of time listening to customers, who told us what they wanted our software to do; that's the innovation demand side, and it gave us crucial information that shaped our research efforts.

Expanding the Supply of Innovation

The work in this first phase is classic research and development, where great scientists and engineers dream up the technologies we need. Although we have a number of cost-competitive low-carbon solutions today, we still don't have all the technologies we need to get to zero emissions globally. I mentioned the most important ones we still need in chapters 4 through 9; here's the list again for quick reference (you can put the words "cheap enough for middle-income countries to buy" in every item on the list):

Technologies needed

Hydrogen produced without emitting carbon	Nuclear fusion
Grid-scale electricity storage that can last a full season	Carbon capture (both direct air capture and point capture)
Electrofuels	Underground electricity transmission
Advanced biofuels	Zero-carbon plastics
Zero-carbon cement	Geothermal energy
Zero-carbon steel	Pumped hydro
Plant- and cell-based meat and dairy	Thermal storage
Zero-carbon fertilizer	Drought- and flood-tolerant food crops
Next-generation nuclear fission	Zero-carbon alternatives to palm oil
	Coolants that don't contain F-gases

To get these technologies ready soon enough to make a difference, governments need to do the following:

1. Quintuple clean energy and climate-related R&D over the next decade. Direct public investment in research and development is one of the most important things we can do to fight climate change, but governments aren't doing nearly enough of it. In total, government funding for clean energy R&D amounts to about $22 billion per year, only around 0.02 percent of the global economy. Americans spend more than that on gasoline in a single month. The United States, which is by far the largest investor in clean energy research, spends only about $7 billion per year.

How much should we spend? I think the National Institutes of Health (NIH) is a good comparison. The NIH, with a budget of about $37 billion a year, has developed lifesaving drugs and treatments that Americans—and people around the world—rely on every day. It's a great model, and an example of the ambition we need for climate change. And although quintupling an R&D budget sounds like a lot of money, it pales in comparison to the size of the challenge—and it's a powerful indicator of just how seriously a government takes the problem.

2. Make bigger bets on high-risk, high-reward R&D projects.
It's not just a question of how much governments spend. What they spend it on matters too.

Governments have been burned by investing in clean energy before (look up "Solyndra scandal" if you need a reminder), and policy makers understandably don't want to look as if they are wasting taxpayers' money. But this fear of failure makes R&D portfolios shortsighted. They tend to skew toward safer investments that could and should be funded by the private sector. The real value of government leadership in R&D is that it can take chances on bold ideas that might fail or might not pay off right away. This is especially true of scientific enterprises that remain too risky for the private sector to pursue for the reasons I touched on in chapter 10.

To see what happens when governments make a big bet the right way, consider the Human Genome Project (HGP). Designed to map the complete set of human genes and make the results available to the public, it was a landmark research project led by the U.S. Department of Energy and the National Institutes of Health, with partners in the U.K., France, Germany, Japan, and China. The project took 13 years and billions of dollars, but it has pointed the way to new tests or treatments for dozens of genetic conditions, including inherited colon cancer, Alzheimer's disease, and familial breast cancer. An independent study of the HGP found that every $1 invested by the federal government in the project generated $141 in returns to the U.S. economy.

By the same token, we need governments to commit to funding mega-scale projects (in the range of hundreds of millions or billions of dollars) that can advance the science of clean energy—especially in the areas I listed above. And they need to commit to funding them for the long haul so that researchers know they'll have a steady flow of support for years to come.

3. Match R&D with our greatest needs. There's a practical distinction between blue-sky research into novel scientific concepts (also called basic research) and efforts to take scientific discoveries and make them useful (what's known as applied or translational research). Although they're different things, it's a mistake to think—as some purists do—that basic science shouldn't be tainted by considering how it might lead to a useful commercial product. Some of the best inventions have emerged when scientists start their research with an end use in mind; Louis Pasteur's work on microbiology, for example, led to vaccines and pasteurization. We need more government programs that integrate basic and applied research in the areas where we most need breakthroughs.

The U.S. Department of Energy's SunShot Initiative is a good example of how this can work. In 2011, the program's leaders set a goal of driving down the cost of solar energy to $0.06 per kilowatt-hour within the decade. They focused on early-stage R&D, but they also encouraged private companies, universities, and national laboratories to concentrate on efforts like lowering the cost of solar-power systems, removing bureaucratic barriers, and making it cheaper to finance a solar-power system. Thanks to this integrated approach, SunShot met its goal in 2017, three years ahead of schedule.

4. Work with industry from the beginning. Another artificial distinction I've run into is the idea that early-stage innovation is for governments and later-stage innovation is for industries. This just isn't how it works in reality—especially when it comes to the kinds of tough technical challenges we have in energy, where the most important measure of success for any idea is the ability to reach national or even global scale. Partnerships at an early stage will bring in the people who know how to do that. Governments and industry will need to work together to overcome barriers and speed up the innovation cycle. Businesses can help prototype new technologies, provide insight into the marketplace, and co-invest in projects. And, of course, they're the ones who will commercialize technology, so it makes sense to bring them in early.

Accelerating the Demand for Innovation

The demand side is a little more complicated than the supply piece. It actually involves two steps: the proof phase, and the scale-up phase.

After an approach has been tested in the lab, it needs to be proven in the market. In the tech world, this proof phase is quick and cheap; it doesn't take long to demonstrate whether a new smartphone model works and will appeal to customers. But in energy, it's much harder and more expensive.

You have to find out whether the idea that worked in the laboratory still works under real-world conditions. (Maybe the agricultural waste you want to turn into a biofuel is much wetter than the stuff you used in the lab and therefore doesn't produce as much energy as you expected.) You also have to drive down the cost and risks of early adoption, develop supply chains, test your business model, and help consumers get comfortable with the new technology. Ideas currently in the proof phase include low-carbon cement, next-generation nuclear fission, carbon capture and sequestration, offshore wind, cellulosic ethanol (a type of advanced biofuel), and meat alternatives.

The proof phase is a valley of death, a place where good ideas go to die. Often, the risks that come with testing new products and introducing them in the market are simply too great. Investors get scared off. This is particularly true for low-carbon technologies, which can require a lot of capital to get going and may require consumers to change their behavior pretty substantially.

Governments (as well as big companies) can help energy start-ups make it out of the valley alive because they're massive consumers. If they prioritize buying green, they'll help bring more products to market by creating certainty and reducing costs.

Use procurement power. Governments at all levels—national, state, and local—buy enormous amounts of fuel, cement, and steel.

They build and operate planes and trucks and cars, and they consume gigawatts' worth of electricity. This puts them in the perfect position to drive emerging technologies into the market at relatively low cost—especially if you factor in the social benefits of bringing these technologies to scale. Defense departments can commit to buying some low-carbon liquid fuels for planes and ships. State governments can use low-emissions cement and steel in construction projects. Utilities can invest in long-duration storage.

Every bureaucrat who makes purchasing decisions should have an incentive to look for green products, understanding how to figure in the cost of the externalities we talked about in chapter 10.

By the way, this isn't a particularly new idea. It's how the internet took off in the early days: There was public R&D funding, of course, but also a committed buyer—the U.S. government—waiting on the other end.

Create incentives that lower costs and reduce risk. In addition to buying things themselves, governments can give the private sector various incentives to go green. Tax credits, loan guarantees, and other tools can help reduce the Green Premiums and drive demand for new technologies. Because many of these products will be expensive for some time to come, prospective buyers will need access to long-term financing, as well as the confidence that comes from consistent and predictable government policies.

Governments can play a huge role by adopting zero-carbon policies and shaping the way markets attract money for these projects. A few principles: Government policies should be *technology neutral* (benefiting any solutions that reduce emissions, rather than a few favored ones), *predictable* (as opposed to regularly expiring and then getting extended, as happens frequently now), and *flexible* (so that many different companies and investors can take advantage of them, not just those with large federal tax bills).

Build the infrastructure that will get new technologies to market. Even cost-competitive low-carbon technologies won't be able to gain market share if the infrastructure isn't in place to get them to market in the first place. Governments at all levels need to help get that infrastructure built. This includes transmission lines for wind and solar, charging stations for electric vehicles, and pipelines for captured carbon dioxide and hydrogen.

Change the rules so new technologies can compete. Once the infrastructure is built, we'll need new market rules that allow the new technologies to be competitive. Electricity markets that were designed around 20th-century technologies often put 21st-century technologies at a disadvantage. For example, in most markets, utilities that invest in long-duration storage aren't appropriately compensated for the value they're providing to the grid. Regulations make it hard to use more advanced biofuels in cars and trucks. And, as I mentioned in chapter 10, some new forms of low-carbon concrete can't compete because of outdated government rules.

So far in this chapter, I've been covering the development phase—policies that can spark the creation and adoption of energy breakthroughs. Now let's turn to the scale-up phase—rapid, large-scale deployment. You can only reach this stage once the cost is low enough, your supply chains and business models are well developed, and consumers have shown that they'll buy what you're selling. Onshore wind, solar power, and electric vehicles are all in the scale-up phase.

But scaling them up won't be easy. We need to more than triple the amount of power in just a few decades, with the majority of the new electrons coming from wind, solar, and other forms of clean energy. We need to be adopting electric vehicles as fast as we bought clothes dryers and color TVs when those became available. We need to transform the way we make and grow things while continuing to deliver the roads, bridges, and food we all rely on.

Luckily, as I mentioned in chapter 10, we're no strangers to scaling up energy technologies. We drove rural electrification and expanded the domestic production of fossil fuels by tying policy and innovation together. You might consider some of those policies—like various tax advantages for oil companies—subsidies for fossil fuels, but they're really just a tool for deploying a technology we thought was valuable. Remember that until the late 1970s—when the concept of climate change first entered the national debate—it was widely accepted that the best way to raise the quality of life and spread economic development was to expand the use of fossil fuels. Now we can take the lessons we learned from the purposeful growth of fossil fuels and apply them to clean energy.

What does that mean in practice?

Put a price on carbon. Whether it's a carbon tax or a cap-and-trade system where companies can buy and sell the right to emit carbon, putting a price on emissions is one of the most important things we can do to eliminate Green Premiums.

In the near term, the value of a carbon price is that by raising the cost of fossil fuels, it tells the market that there will be extra costs associated with products that emit greenhouse gas emissions. Where the revenue from this carbon price goes is not as important as the market signal sent by the price itself. Many economists argue that the money can be returned to consumers or businesses to cover the resulting increase in energy prices, though there's also a strong argument that it should go to R&D and other incentives to help solve climate change.

In the longer term, as we get closer to net-zero emissions, the carbon price could be set at the cost of direct air capture, and the revenues could be used to pay for pulling carbon out of the air.

Although it would be a fundamental shift in the way we think about pricing goods, the concept of a carbon price has broad acceptance among economists from many schools of thought and across the political spectrum. Getting it right is going to be technically and politically hard, in the United States and around the world. Will people be willing to pay that much more for their gasoline and every other product in their lives that involves greenhouse gas emissions, which is pretty much all of them? I'm not going to prescribe what the solution should look like, but the core objective is to make sure everyone pays the true cost of their emissions.

Clean electricity standards. Twenty-nine U.S. states and the European Union have adopted a type of performance standard called a renewable portfolio standard. The idea here is to require electrical utilities to get a certain percentage of their electricity from renewable sources. These are flexible, market-based mechanisms; for example, utilities with access to more renewable resources can sell credits to those with fewer. But there's a problem with the way this approach is carried out today: It limits utilities to using only certain approved low-carbon technologies (wind, solar, geothermal, sometimes hydro), and it excludes options like nuclear power and carbon capture. That effectively raises the overall cost of lowering emissions.

Clean electricity standards, which a growing number of states are now looking to adopt, are a better way to go. Rather than emphasizing renewable sources in particular, they allow any clean energy technology—including nuclear and carbon capture—to count toward meeting the standard. It's a flexible, cost-effective approach.

Clean fuel standards. This idea of flexible performance standard can be applied to other sectors too, to reduce the emissions from cars and buildings as well as power plants. For example, a clean fuel standard applied to the transportation sector would accelerate deployment of electric vehicles, advanced biofuels, electrofuels, and other low-carbon solutions. As with a clean electricity standard, it would be technology neutral, and regulated entities could be allowed to trade credits, both of which lower the cost to consumers. California has created a model for this with its Low Carbon Fuel Standard. At the national level, the United States has the basis for such a policy with the Renewable Fuel Standard, which can be reformed to address the limitations I mentioned in chapter 10 and expanded to cover other low-carbon solutions (including electricity and electrofuels). This would make it a powerful tool in addressing climate change. The EU's Renewable Energy Directive provides a similar opportunity in Europe.

Clean product standards. Performance standards can also help accelerate the deployment of low-emissions cement, steel, plastics, and other carbon-intensive products. Governments can start the process by setting standards in their procurement programs and by creating labeling programs that give all buyers information about how "clean" different suppliers are. Then we can expand these to standards covering all carbon-intensive goods sold in a market, not just whatever's being bought by governments. Imported goods would have to qualify too, which would address countries' concerns that lowering emissions from their manufacturing sectors will make their products more expensive and put them at a competitive disadvantage.

Out with the old. In addition to rolling out new technology as fast as possible, governments will need to retire inefficient, fossil-fueled equipment—whether power plants or automobiles—faster than they might otherwise. It costs a lot to build power plants, and the energy they produce is only cheap if you can spread the cost of construction over their useful life span. As a result, utility companies and their regulatory agencies are loath to shut down a perfectly good operating plant that may have decades of useful life ahead of it. Policy-based incentives, through either the tax code or utility regulation, can accelerate this process.

Who's on First?

No single government body could fully implement a plan like the one I've outlined; the decision-making authority is simply too dispersed. We'll need action at all levels of government, from local transportation planners to national legislatures and environmental regulators.

The exact mix will vary from one country to another, but I'll touch on a few common themes that are true in most places today.

Local governments play an important role in determining how buildings are constructed and what kinds of energy they use, whether buses and police cars run on electricity, whether there's a charging infrastructure for electric vehicles, and how waste gets managed.

Most state or provincial governments have a central role in regulating electricity, planning infrastructure like roads and bridges, and selecting the materials that go into these projects.

National governments generally have authority over activities that cross state or international borders, so they write the rules that shape electricity markets, adopt pollution regulations, and create standards for vehicles and fuels. They also have enormous procurement power, are the primary source of fiscal incentives, and usually fund more public R&D than any other level of government.

In short, every national government needs to do three things.

First, make it a goal to get to zero—by 2050 for rich countries, and as soon after 2050 as possible for middle-income countries.

Second, develop specific plans for meeting those goals. To get to zero by 2050, we'll need to have the policy and market structures in place by 2030.

And third, any country that's in a position to fund research needs to make sure it's on track to make clean energy so cheap—to reduce the Green Premiums so much—that middle-income countries will be able to get to zero.

To show you how it can all work together, here's what a whole-of-government approach to accelerating innovation could look like in the United States.

Federal Government

The U.S. government does more to drive the supply of energy innovation than anyone else. It's the biggest funder and performer of energy research and development, with 12 different federal agencies involved in research (the Department of Energy has by far the largest share). It has all sorts of tools for managing the direction and pace of energy R&D: research grants, loan programs, tax incentives, laboratory facilities, pilot programs, public-private partnerships, and more.

The federal government also plays a central role in driving the demand for green products and policies. It helps fund roads and bridges built by state and local governments, regulates cross-state infrastructure like transmission lines, pipelines, and highways, and helps set the rules for multistate electricity and fuel markets. And it collects most tax revenue, which means that federal financial incentives will be the most effective at driving change.

When it comes to scaling up new technologies, the federal government plays the largest role of anyone. It regulates interstate commerce and has primary authority over international trade and investment policy, meaning we'll need federal policies to reduce any sources of emissions that cross state lines or international borders. (According to *The Economist*—one of my favorite magazines—U.S. emissions would be about 8 percent higher if you included all the products that Americans consume but are made elsewhere. Britain's would be about 40 percent higher.) Although carbon pricing, clean electricity standards, clean fuel standards, and clean product standards should all be adopted at a state level, they'll be more effective if they're implemented across the country.

In practice, that means Congress needs to provide funding for R&D, government procurement, and developing infrastructure, and it needs to create, modify, or extend financial incentives for green policies and products.

Meanwhile, in the executive branch, the Department of Energy does in-house research and funds other work as well; it would play a central role in implementing a federal clean electricity standard. The Environmental Protection Agency would be charged with designing and implementing an expanded clean fuel standard. The Federal Energy Regulatory Commission, which oversees wholesale electricity markets and interstate transmission and pipeline projects, would need to regulate the infrastructure and market elements of a plan.

The list goes on. The Department of Agriculture does key work on land use and agricultural emissions; the Department of Defense buys advanced low-emissions fuels and materials; the National Science Foundation funds research; the Department of Transportation helps fund roads and bridges; and so on.

Finally, there's the matter of how we finance the work it'll take to get to zero. We can't know with any precision how much getting to zero will cost over time—it will depend on the success and speed of innovation and the effectiveness of deployment—but we know that it will require massive investment.

The United States is lucky to have mature and creative capital markets that can grab great ideas and get them developed and deployed quickly; I've suggested ways that the federal government can help move those markets in the right direction and partner with the private sector in new ways. Other countries—China, India, and many European nations, for example—don't have private markets that are as strong, but they can still make big public investments for climate change. And multilateral banks, like the World Bank and development banks in Asia, Africa, and Europe, are also looking to get more involved.

Two things are clear. First, the amount of money invested in getting to zero, and adapting to the damage that we know is coming, will need to ramp up dramatically and for the long haul. To me, this means that governments and multilateral banks will need to find much better ways to tap private capital. Their coffers aren't big enough to do this on their own.

Second, the time frames for climate investment are long, and the risks are high. So the public sector should be using its financial strength to lengthen the investment horizon—reflecting the fact that returns may not come for many years—and reduce the risk of these investments. It'll be tricky to mix public and private money on such a large scale, but it's essential. We need our best minds in finance working on this problem.

State Governments

In America, many states are leading the way. Twenty-four states and Puerto Rico have joined the bipartisan U.S. Climate Alliance, committing to meet the Paris Agreement goals of reducing emissions by at least 26 percent by 2025. Although that's not nearly enough to reduce nationwide emissions as much as we need to, it's not tilting at windmills either. States can play a crucial role in demonstrating innovative technologies and policies, such as using their utilities and road construction projects to drive technologies like long-duration storage and low-emissions cement into the market.

States can also test policies like carbon pricing, clean electricity standards, and clean fuel standards before we implement them across the country. And they can join together in regional alliances, the way California and other western states are looking at joining up their grids and as some states in the Northeast have done with a cap-and-trade program to lower emissions. The U.S. Climate Alliance and the cities that have aligned with it represent more than

60 percent of the U.S. economy, which means they have a phenomenal ability to create markets and show how we can get new ideas to scale.

State legislatures would be responsible for adopting state-level carbon-pricing systems, clean energy standards, and clean fuel standards. They would also direct state agencies and public utility or service commissions to change their procurement policies so they prioritize advanced low-emissions technologies.

State agencies are responsible for meeting goals set by the legislature and by the governor. They oversee energy efficiency and buildings-related policy, manage state transport-related policy and investment, enforce pollution standards, and regulate agriculture and other uses of land.

In the unlikely event that someone runs up to you and demands, "What's the most obscure agency with an underappreciated impact on climate change?" you could do worse than say, "My state's public utility commission," or "My state's public service commission." (The name varies from state to state.) Most people have never heard of PUCs or PSCs, but they're actually responsible for many of the regulations related to electricity in the United States. For example, they approve investment plans proposed by electric utilities and determine the price that consumers pay for electricity. And they'll become only more important as we meet more of our energy needs with electricity.

Local Governments

Mayors across the United States and around the world are committing to reduce emissions. Twelve major American cities have set a goal of being carbon neutral by 2050, and more than 300 have pledged to meet the goals of the Paris Agreement.

Cities don't have as much influence on emissions as state and federal governments, but they're far from powerless. Although they can't set their own vehicle emissions standards, for example, they can buy electric buses, fund more charging stations for electric vehicles, use zoning laws to increase density so people travel less between work and home, and potentially restrict access to their roads by fossil-fuel-powered vehicles. They can also establish green building policies, electrify their vehicle fleets, and set procurement guidelines and performance standards for municipally owned buildings.

And some cities—Seattle, Nashville, and Austin, for example—own the local utility company, giving them oversight over whether they get their electricity from clean sources. Cities like these can also allow the building of clean energy projects on city land.

City councils can take action similar to that of state legislatures and the U.S. Congress, funding climate policy priorities and requiring local government agencies to take action.

Local agencies, like their state and federal counterparts, oversee different policy priorities. Building departments enforce efficiency requirements; transit agencies can go electric and influence the materials used in roads and bridges; waste management agencies operate large vehicle fleets and have influence over emissions from landfills.

Back to the federal level for one last point: how rich countries can help eliminate the free-rider problem.

There's no way to sugarcoat the fact that getting to zero won't come for free. We have to invest more money in research, and we need policies that drive the markets toward clean energy products that are, right now, more expensive than their greenhouse-gas-emitting counterparts.

But it's hard to impose higher costs now in exchange for a better climate later. The Green Premiums give countries, and especially middle- and low-income countries, a major incentive to resist cutting their emissions. We've already seen example after example around the world—Canada, the Philippines, Brazil, Australia, France, and others—in which the public makes it clear with their votes and their voices that they don't want to pay more for gasoline, heating oil, and other basics.

The problem is not that people in these countries want the climate to get hotter. The problem is that they're worried about how much the solutions will cost them.

So how do we solve the free-rider problem?

It helps to set ambitious goals and commit to meeting them, the way countries around the world did with the 2015 Paris Agreement. It's easy to mock international agreements, but they're part of how progress happens: If you like having an ozone layer, you can thank an international agreement called the Montreal Protocol.

Once these goals are set, forums like the COP 21 are where countries get together to report on their progress and share what's working. And they serve as a mechanism for pressing national governments to do their part. When the world's governments agree that there's value in reducing emissions, it becomes harder—though far from impossible, as we have seen—to be the outlier who says, "I don't care. I'm going to keep emitting greenhouse gases."

What about those who refuse to go along? It's notoriously difficult to hold a country accountable for something like its carbon emissions. But it's not out of the question. For example, governments that adopt a price on carbon can create what's called a border adjustment—making sure that the carbon price on some product is paid whether that product was made domestically or imported from somewhere else. (They'd need to make allowances for products from low-income countries where the priority is to drive economic growth, not to reduce their already very low carbon emissions.)

And even countries without a carbon tax can make it clear that they won't make trade agreements and enter multilateral partnerships with anyone who hasn't made it a priority to reduce greenhouse gases and adopted the policies to accomplish it (again, with allowances for low-income countries). In essence, governments can say to each other, "If you want to do business with us, you'll have to take climate change seriously."

Finally, and in my view most important, we have to lower the Green Premiums. It's the only way to make it easier for middle- and low-income countries to reduce their emissions and eventually get to zero, and it will happen only if rich countries—especially the United States, Japan, and European nations—take the lead. After all, that's where much of the world's innovation happens.

And—this is a really important point—*lowering the Green Premiums that the world pays is not charity*. Countries like the United States shouldn't see investing in clean energy R&D as just a favor to the rest of the world. They should also see it as an opportunity to make scientific breakthroughs that will give birth to new industries composed of major new companies, creating jobs and reducing emissions at the same time.

Think about all the good that comes from medical research funded by the National Institutes of Health. The NIH publishes its results so scientists around the world can benefit from the work, but its funding also builds up capacity in American universities that are, in turn, connected to both start-ups and big companies. The result: an American export—advanced medical expertise—that creates a lot of high-paying jobs at home and saves lives around the world.

It's a similar story in technology, where early investments by the Department of Defense led to the creation of the internet and the microchips that powered the personal computer revolution.

And the same thing can happen in clean energy. There are markets worth billions of dollars waiting for someone to invent low-cost, zero-carbon cement or steel, or a net-zero liquid fuel. As I've tried to show, making these breakthroughs and getting them to scale will be hard, but the opportunities are so big that it's worth getting out in front of the rest of the world. Someone will invent these technologies. It's just a question of who and how soon.

There's a lot that individuals can do, from the local level to the national level, to accelerate this agenda. We'll cover that in the next and final chapter.

CHAPTER 12

WHAT EACH OF US CAN DO

I t's easy to feel powerless in the face of a problem as big as climate change. But you're not powerless. And you don't have to be a politician or a philanthropist to make a difference. You have influence as a citizen, a consumer, and an employee or employer.

As a Citizen

When you ask yourself what you can do to limit climate change, it's natural to think of things like driving an electric car or eating less meat. This sort of personal action is important for the signals it sends to the marketplace—see the next section for more on that point—but the bulk of our emissions comes from the larger systems in which we live our daily lives.

When somebody wants toast for breakfast, we need to make sure there's a system in place that can deliver the bread, the toaster, and the electricity to run the toaster without adding greenhouse gases to the atmosphere. We aren't going to solve the climate problem by telling people not to eat toast.

But putting this new energy system in place requires concerted political action. That's why engaging in the political process is the most important single step that people from every walk of life can take to help avoid a climate disaster.

In my own meetings with politicians, I've found that it helps to remember that climate change isn't the only thing on their plate. Government leaders are also thinking about education, jobs, health care, foreign policy, and more recently, COVID-19. And they should: All those things demand attention.

But policy makers can take on only so many problems at once. And they decide what to do, what to prioritize, based on what they're hearing from their constituents.

In other words, elected officials will adopt specific plans for climate change if their voters demand it. Thanks to activists around the world, we don't need to generate demand: Millions of people are already calling for action. What we do need to do, though, is to translate these calls for action into pressure that encourages politicians to make the tough choices and trade-offs necessary to deliver on their promises to reduce emissions.

Whatever other resources you may have, you can always use your voice and your vote to effect change.

Make calls, write letters, attend town halls. What you can help your leaders understand is that it's just as important for them to think about the long-term problem of climate change as it is for them to think about jobs or education or health care.

It may sound old-fashioned, but letters and phone calls to your elected officials can have a real impact. Senators and representatives get frequent reports on what their offices are hearing from constituents. But don't simply say, "Do something about climate change." Know where they stand, ask them questions, and make clear that this is an issue that will help determine how you vote. Demand more funding for clean energy R&D, a clean energy standard, a price on carbon, or any of the other policies from chapter 11.

Look locally as well as nationally. A lot of the relevant decisions are made at the state and local levels by governors, mayors, state legislatures, and city councils—places where individual citizens can have an even bigger impact than at the federal level. In the United States, for example, electricity is primarily regulated by statewide public utility commissions, made up of either elected or appointed commissioners. Know who your representatives are and keep in touch with them.

Run for office. Running for the U.S. Congress is a tall order. But you don't have to start there. You can run for state or local office, where you'll probably have more impact anyway. We need all the policy smarts, courage, and creativity in public office that we can get.

As a Consumer

The market is ruled by supply and demand, and as a consumer you can have a huge impact on the demand side of the equation. If all of us make individual changes in what we buy and use, it can add up to a lot—as long as we focus on changes that are meaningful. For example, if you can afford to install a smart thermostat to reduce your energy consumption when you're not at home, by all means do it. You'll cut your utility bill and your greenhouse gas emissions.

But reducing your own carbon emissions isn't the most powerful thing you can do. You can also send a signal to the market that people want zero-carbon alternatives and are willing to pay for them. When you pay more for an electric car, a heat pump, or a plant-based burger, you're saying, "There's a market for this stuff. We'll buy it." If enough people send the same signal, companies will respond—quite quickly, in my experience. They'll put more money and time into making low-emissions products, which will drive down the prices of those products, which will help them get adopted in big numbers. It will make investors more confident about funding new companies that are making the breakthroughs that will help us get to zero.

Without that demand signal, the innovations that governments and businesses invest in will stay on the shelf. Or they won't get developed in the first place, because there's no economic incentive to make them.

Here are some specific steps you can take:

Sign up for a green pricing program with your electric utility. Some utility companies allow homes and businesses to pay extra for power from clean sources. In 13 states, utilities are required to offer this option. (You can see whether your state does by checking the Green Pricing Programs map at C2ES—the Center for Climate and Energy Solutions—www.c2es.org/document/green -pricing-programs.) Customers in these programs pay a premium on their electric bill to cover the extra cost of renewable energy, an average of one to two cents per kilowatt-hour, or $9 to $18 a month for the typical American home. When you participate in these programs, you're telling your utility company that you're willing to pay more to address climate change. That's an important market signal.

But what these programs don't do is cancel out emissions or lead to meaningful increases in the amount of renewable power on the grid. Only government policies and increased investments can do that.

Reduce your home's emissions. Depending on how much money and time you can spare, you can replace your incandescent lightbulbs with LEDs, install a smart thermostat, insulate your windows, buy efficient appliances, or replace your heating and cooling system with a heat pump (as long as you live in a climate where they can operate). If you rent your home, you can make the changes within your control—such as replacing lightbulbs—and encourage your landlord to do the rest. If you're building a new home or renovating an old one, you can opt for recycled steel and make the home more efficient by using structural insulated panels, insulating concrete forms, attic or roof radiant barriers, reflective insulation, and foundation insulation.

Buy an electric vehicle. EVs have come a long way in terms of cost and performance. Although they might not be right for everyone (they're not great for lots of long-distance road trips, and charging at home isn't convenient for everyone), they're becoming more affordable for many consumers. This is one of the places where consumer behavior can have a huge impact: If people buy lots of them, companies will make lots of them.

Try a plant-based burger. I'll admit that veggie burgers haven't always tasted great, but the new generation of plant-based protein alternatives is better and closer to the taste and texture of meat than their predecessors. You can find them in many restaurants, grocery stores, and even fast-food joints. Buying these products sends a clear message that making them is a wise investment. In addition, eating a meat substitute (or simply not eating meat) just once or twice a week will cut down on the emissions you're responsible for. The same goes for dairy products.

As an Employee or Employer

As an employee or a shareholder, you can push your company to do its part. Of course, big companies will have the largest impact in many of these areas, but small companies can also do a lot, especially if they work together through organizations like local chambers of commerce.

Some steps are easier than others. The easy things do matter—planting trees to offset emissions, for instance, is a good thing to do for environmental and political reasons. It demonstrates that you care about climate change.

But doing only the easy things won't solve the problem. The private sector will also need to embrace the harder steps.

For one thing, that means accepting more risk—for example, financing projects that might fail, but might turn into a clean-energy breakthrough. Shareholders and board members will have to be willing to share in this risk, making it clear to executives that they'll back smart investments even if they don't ultimately pan out. Companies and their leaders need to be rewarded for making bets that could move us forward on climate change.

Companies can also work with each other to identify and try to solve the toughest climate challenges. That means looking for the biggest Green Premiums and trying to reduce them. If the world's biggest private-sector consumers of materials like steel and cement got together and demanded cleaner substitutes—and committed to investing in the infrastructure needed to make them—it would accelerate research and shift the market in the right direction.

Finally, the private sector can advocate for making these hard choices—for instance, by agreeing to use its resources to develop these markets, and by demanding that governments set up regulatory structures in which new technologies can succeed. Are political leaders focusing on the biggest sources of emissions and the toughest technical challenges? Are they talking about grid-scale energy storage, electrofuels, nuclear fusion, carbon capture, and zero-carbon cement and steel? If not, they're not helping us get on the path to zero emissions by 2050.

Here are some specific steps the private sector can take along these lines:

Set up an internal carbon tax. Some big companies now impose a carbon tax on each of their divisions. These companies aren't paying lip service to reducing emissions. They're helping products get out of the lab and into the market, because the revenue from internal taxes can go directly to activities that reduce the Green Premiums and help create a market for the clean-energy products those firms will need. Employees, investors, and customers can advocate for this approach, giving cover to the executives responsible for implementing it.

Prioritize innovation in low-carbon solutions. Investing in new ideas used to be a point of pride for most industries, but the glory years of corporate R&D are gone. Today, companies in the aerospace, materials, and energy industries spend on average less than 5 percent of their revenue on R&D. (Software companies spend upwards of 15 percent.) Companies should reprioritize their R&D work, particularly on low-carbon innovations, many of which will require long-term commitments. Larger companies can partner with government researchers to bring practical commercial experience to research efforts.

Be an early adopter. Like governments, companies can use the fact that they buy a lot of products to speed up the adoption of new technologies. Among other things, this can involve using electric vehicles for corporate fleets, buying lower-carbon materials to build or renovate company buildings, and committing to use a certain amount of clean electricity. Many companies around the world have already committed to using renewable power for a large part of their operations, including Microsoft, Google, Amazon, and Disney. The shipping company Maersk has said it will cut its net emissions to zero by 2050.

Even if these commitments will be hard to meet, they send important market signals about the value of developing zero-carbon approaches. Innovators see the demand and know they'll have a market ready to buy their products.

Engage in the policy-making process. Companies can't be afraid of working with the government, any more than governments should be afraid of working with companies. Businesses should champion getting to zero and support funding for basic science and applied R&D programs that will get us there. This will be especially important given the decline in corporate R&D over the past several decades.

Connect with government-funded research. Businesses should be advising government R&D programs so that basic and applied research is focused on the ideas that have the best shot of turning into products. (No one knows what is or isn't likely to succeed better than the companies that develop and market products every day.) Joining industry advisory boards and taking part in planning exercises are low-cost ways to inform government R&D programs.

Companies can also help fund R&D through cost-sharing agreements and joint research projects—the kind of public-private collaboration that gave us gas turbines and advanced diesel engines.

Help early-stage innovators get across the valley of death. Many researchers never turn their promising ideas into products because the process would be too risky or too expensive. Established businesses can help by providing access to their testing facilities and providing data like cost metrics. If they want to do more, they can offer fellowships and incubation programs for entrepreneurs, invest in new technology, create business divisions that are specifically focused on low-carbon innovation, and finance new low-emissions projects.

One Last Thought

Unfortunately, the conversation about climate change has become unnecessarily polarized, not to mention clouded by conflicting information and confusing stories. We need to make the debate more thoughtful and constructive, and most of all we need to center it on realistic, specific plans for getting to zero.

I wish there were some magic invention that could steer the conversation in a more productive direction. Of course, no such device exists. Instead, it's up to each of us.

My hope is that we can shift the conversation by sharing the facts with the people in our lives—our family members, friends, and leaders. And not just the facts that tell us why we need to act, but also those that show us the actions that will do the most good. One of my goals in writing this book is to spark more of these conversations.

I also hope we can unite behind plans that bridge political divides. As I've tried to demonstrate, this isn't as naive as it may sound. No one has cornered the market on effective solutions to climate change. Whether you're a believer in the private sector, or government intervention, or activism, or some combination, there's a practical idea you can get behind. As for the ideas you can't support, you may feel compelled to speak out, and that's understandable. But I hope you'll spend more time and energy supporting whatever you're in favor of than opposing whatever you're against.

With the threat of climate change upon us, it can be hard to be hopeful about the future. But as my friend Hans Rosling, the late global health advocate and educator, wrote in his amazing book *Factfulness:* "When we have a fact-based worldview, we can see that the world is not as bad as it seems—and we can see what we have to do to keep making it better."

When we have a fact-based view of climate change, we can see that we have some of the things we need to avoid a climate disaster, but not all of them. We can see what stands in the way of deploying the solutions we have and developing the breakthroughs we need. And we can see all the work we must do to overcome those hurdles.

I'm an optimist because I know what technology can accomplish and because I know what *people* can accomplish. I'm profoundly inspired by all the passion I see, especially among young people, for solving this problem. If we keep our eye on the big goal—getting to zero—and we make serious plans to achieve that goal, we can avoid a disaster. We can keep the climate bearable for everyone, help hundreds of millions of poor people make the most of their lives, and preserve the planet for generations to come.

CLIMATE CHANGE AND COVID-19

I finished working on this book at the end of the most tumultuous year in recent memory. As I write this afterword in November 2020, COVID-19 has killed more than 1.4 million people around the world and is entering another wave of cases and deaths. The pandemic has changed the way we work, live, and socialize.

At the same time, 2020 also brought new reasons to be hopeful about climate change. With the election of Joe Biden as president, the United States is poised to resume a leading role on the issue. China committed to the ambitious goal of being carbon neutral by 2060. In 2021, the United Nations will gather in Scotland for another major summit on climate change. Of course, none of this guarantees that we'll make progress, but the opportunities are there.

I expect to spend much of my time in 2021 talking with leaders around the world about both climate change and COVID-19. I will make the case to them that many of the lessons from the pandemic—and the values and principles that guide our approach to it—apply just as well to climate change. At the risk of repeating myself from earlier in this book, I'll summarize them here.

First, we need international cooperation. The phrase "we have to work together" is easy to dismiss as a cliché, but it's true. When governments, researchers, and pharmaceutical companies worked together on COVID-19, the world made remarkable progress—for example, developing and testing vaccines in record time. And when we didn't learn from each other and instead demonized other countries, or refused to accept that masks and social distancing slow the spread of the virus, we extended the misery.

The same is true for climate change. If rich countries worry only about lowering their own emissions and don't work to make clean technologies practical for everyone, we'll never get to zero. In that sense, helping others is not just an act of altruism, it's also in our self-interest. We all have reasons to get to zero and help others do it, too. Temperatures will not stop rising in Texas unless emissions stop rising in India.

Second, we need to let science—actually, many different sciences—guide our efforts. In the case of COVID-19, we are looking to biology, virology, and pharmacology, as well as political science and economics—after all, deciding how to distribute vaccines equitably is an inherently political act. And just as epidemiology tells us about the risks of COVID-19 but not how to stop it, climate science tells us why we need to change course but not how to do it. For that, we must draw on engineering, physics, environmental science, economics, and more.

Third, our solutions should meet the needs of the people who are hardest hit. With COVID-19, the people who suffer most are the ones who have the fewest options—working from home, for example, or taking time off to care for themselves or their loved ones. And most of them are people of color and lower-income people.

In the United States, Black people and Latinx people are disproportionately likely to contract the coronavirus and to die from it. Black and Latinx students are also less likely to be able to attend school online than their white peers. Among recipients of Medicare, the COVID-19 death rate is four times higher for those who are poor. Closing these gaps will be key to controlling the virus in the United States.

Globally, COVID-19 has undone decades of progress on poverty and disease. As governments moved to deal with the pandemic, they had to pull people and money away from other priorities, including vaccination programs. A study by the Institute for Health Metrics and Evaluation found that in 2020, vaccination rates dropped to levels last seen in the 1990s. We lost 25 years of progress in about 25 weeks.

Rich nations, already generous in their giving for global health, will need to be even more generous to make up for this loss. The more they invest in strengthening health systems around the world, the more prepared we will be for the next pandemic.

In the same way, we need to plan for a just transition to a zero-emissions future. As I argued in chapter 9, people in poor countries need help adjusting to a warmer world. And wealthier countries will need to acknowledge that the energy transition will be disruptive for the communities that rely on today's energy systems: the places where coal mining is the main industry, where cement is made, steel is smelted, or cars are manufactured. In addition, many people have jobs that indirectly rely on those industries—when there is less coal and fuel to move around, there will be fewer jobs for truck drivers and railroad workers. A significant portion of the working class economy will be affected, and there should be a transition plan in place for those communities.

Finally, we can do the things that will both rescue economies from the COVID disaster and spark innovation to avoid a climate disaster. By investing in clean-energy research and development—R&D—governments can promote economic recovery that also helps reduce emissions. Although it's true that R&D spending has its biggest impact over the long term, there's also an immediate impact: This money creates jobs quickly. In 2018, the U.S. government's investment in all sectors of research and development directly and indirectly supported more than 1.6 million jobs, producing $126 billion in income for workers and $39 billion in federal and state tax revenue.

R&D isn't the only area where economic growth is connected to zero-carbon innovation. Governments can also help clean-energy companies grow by adopting policies that reduce the Green Premiums and make it easier for green products to compete with their fossil-based competitors. And they can use funding from their COVID relief packages for things like expanding the use of renewables and building integrated electricity grids.

The year 2020 was a huge and tragic setback. But I am optimistic that we will get COVID-19 under control in 2021. And I'm optimistic that we'll make real progress on climate change—because the world is more committed to solving this problem than it has ever been.

When the global economy went into severe recession in 2008, public support for action on climate change plummeted. People just couldn't see how we could respond to both crises at the same time.

This time is different. Even though the pandemic has wrecked the global economy, support for action on climate change is just as high as it was in 2019. Our emissions, it seems, are no longer a problem that we're willing to kick down the road.

The question now is this: What should we do with this momentum? To me, the answer is clear. We should spend the next decade focusing on the technologies, policies, and market structures that will put us on the path to eliminating greenhouse gases by 2050. It's hard to think of a better response to a miserable 2020 than spending the next ten years dedicating ourselves to this ambitious goal.

I want to thank the people at Gates Ventures and Breakthrough Energy who helped make *How to Avoid a Climate Disaster* possible.

Josh Daniel is an invaluable writing partner. He helped me express the complexities of climate change and clean energy as simply and clearly as possible. If this book is as effective as I hope it will be, it is largely due to Josh's skill.

I wrote this book because I want to encourage the world to adopt effective plans for dealing with climate change. In that effort, I could not have better partners than Jonah Goldman and his team, including Robin Millican, Mike Boots, and Lauren Nevin. They provide me with essential advice on climate policy and on strategies to make sure that the ideas in this book will have an impact.

Ian Saunders led the creative and production process with all the ingenuity I've come to count on from him. Anu Horsman and Brent Christofferson designed the charts—with expert help from Beyond Words—and picked the photographs that help bring this book to life.

Bridgitt Arnold and Andy Cook led the promotional effort.

And Larry Cohen managed all this work with his usual calm and wisdom.

The team at the Rhodium Group, led by Trevor Houser and Kate

Larsen, was extraordinarily helpful. Their research and advice are reflected throughout this book.

Thanks also to everyone on the board of Breakthrough Energy Ventures: Mukesh Ambani, John Arnold, John Doerr, Rodi Guidero, Abby Johnson, Vinod Khosla, Jack Ma, Hasso Plattner, Carmichael Roberts, and Eric Toone.

Jabe Blumenthal and Karen Fries are the two former Microsoft colleagues who organized my first learning session on climate change in 2006. In that session, they introduced me to two climate scientists, Ken Caldeira—then at the Carnegie Institution for Science—and David Keith of the Harvard University Center for the Environment. Since then, I've had countless conversations with Ken and David that have shaped my thinking.

Ken and a team of his postdoctoral fellows—Candise Henry, Rebecca Peer, and Tyler Ruggles—pored over the manuscript line by line to check for factual mistakes. I'm thankful for their meticulous work. Any remaining errors are my responsibility.

The late David MacKay of Cambridge University inspired me with his wit and insights. I recommend his phenomenal book *Sustainable Energy—Without the Hot Air* to anyone who wants to dig deeper into the subject of energy and climate change.

Vaclav Smil, a professor emeritus at the University of Manitoba, is one of the finest systems thinkers I have ever met. His influence on this book is particularly evident in the passages on the history of energy transitions, and in the errors he helped me avoid.

I've been lucky enough to get to meet—and learn from—a number of knowledgeable people over the years. Thanks to Senator Lamar Alexander, Josh Bolten, Carol Browner, Steven Chu, Arun Majumdar, Ernest Moniz, Senator Lisa Murkowski, Henry Paulson, and John Podesta for being so generous with their time.

Nathan Myhrvold gave me thoughtful feedback on an early draft. Nathan never hesitates to tell me what he really thinks, a quality I always appreciate, even when I don't take his advice.

Other friends and colleagues kindly took the time to read the manuscript and offer their feedback, including Warren Buffett, Sheila Gulati, Charlotte Guyman, Geoff Lamb, Brad Smith, Marc St. John, Mark Suzman, and Lowell Wood.

I want to thank the rest of the team at Breakthrough Energy: Meghan Bader, Julie Barger, Adam Barnes, Farah Benahmed, Ken Caldeira, Saad Chaudhry, Jay Dessy, Gail Easley, Ben Gaddy, Ashley Grosh, Jon Hagg, Conor Hand, Aliya Haq, Victoria Hunt, Anna Hurlimann, Krzysztof Ignaciuk, Kamilah Jenkins, Christie Jones, Casey Leiber, Yifan Li, Dan Livengood, Jennifer Maes, Lidya Makonnen, Maria Martinez, Ann Mettler, Trisha Miller, Kaspar Mueller, Daniel Muldrew, Philipp Offenberg, Daniel Olsen, Merrielle Ondreicka, Julia Reinaud, Ben Rouillé d'Orfeuil, Dhileep Sivam, Jim VandePutte, Demaris Webster, Bainan Xia, Yixing Xu, and Allison Zelman.

I'm grateful for all the support I get from the team at Gates Ventures. Thanks to Katherine Augustin, Laura Ayers, Becky Bartlein, Sharon Bergquist, Lisa Bishop, Aubree Bogdonovich, Niranjan Bose, Hillary Bounds, Bradley Castaneda, Quinn Cornelius, Zephira Davis, Prarthna Desai, Pia Dierking, Gregg Eskenazi, Sarah Fosmo, Josh Friedman, Joanna Fuller, Meghan Groob, Rodi Guidero, Rob Guth, Diane Henson, Tony Hoelscher, Mina Hogan, Margaret Holsinger, Jeff Huston, Tricia Jester, Lauren Jiloty, Chloe Johnson, Goutham Kandru, Liesel Kiel, Meredith Kimball, Todd Krahenbuhl, Jen Krajicek, Geoff Lamb, Jen Langston, Jordyn Lerum, Jacob Limestall, Abbey Loos, Jennie Lyman, Mike Maguire, Kristina Malzbender, Greg Martinez, Nicole MacDougall, Kim McGee, Emma McHugh, Kerry McNellis, Joe Michaels, Craig Miller, Ray Minchew, Valerie Morones, John Murphy, Dillon Mydland, Kyle Nettelbladt, Paul Nevin, Patrick Owens, Hannah Palko, Mukta Phatak, David Phillips, Tony Pound, Bob Regan, Kate Reizner, Oliver Rothschild, Katie Rupp, Maheen Sahoo, Alicia Salmond, Brian Sanders, KJ Sherman, Kevin Smallwood, Jacqueline Smith, Steve Springmeyer, Rachel

Strege, Khiota Therrien, Caroline Tilden, Sean Williams, Sunrise Swanson Williams, Yasmin Wazir, Cailin Wyatt, Mariah Young, and Naomi Zukor.

I'd like to thank the team at Knopf. Bob Gottlieb's early support for this book helped make it happen. Everything you've heard about his brilliant editing is true. Katherine Hourigan shepherded this book through every phase of editing and production with skill and grace. Thanks also to the late Sonny Mehta, Reagan Arthur, Maya Mavjee, Tony Chirico, Andy Hughes, Paul Bogaards, Chris Gillespie, Lydia Buechler, Mike Collica, John Gall, Suzanne Smith, Serena Lehman, Kate Hughes, Anne Achenbaum, Jessica Purcell, Julianne Clancy, and Elizabeth Bernard. And thanks to Lizzie Gottlieb for introducing this project to her father.

Finally, I want to thank Melinda, Jenn, Rory, and Phoebe; my sisters, Kristi and Libby; and my dad, Bill Gates Sr., who passed away during the writing of this book. I could not ask for a more loving and supportive family.

Introduction: 52 Billion to Zero

5   Photo: James Iroha.

6   Figure: Income and energy use go hand in hand: This graph uses data from the World Bank World Development Indicators, which is licensed under CC BY 4.0 (https://www.creativecommons.org/licenses/by/4.0) and available at https://data.worldbank.org/. Income measured as gross domestic product (GDP) per capita in 2014, based on purchasing power parity (PPP), in current international dollars. Energy use measured in kilograms of oil equivalent per capita in 2014, based on IEA data from the World Bank World Development Indicators. All rights reserved; as modified by Gates Ventures, LLC.

12   Photo: Launching Mission Innovation: From left to right (titles were current at the time of the event in 2015): Wan Gang, Minister of Science and Technology (China); Ali Al-Naimi, Minister of Petroleum and Mineral Resources (Saudi Arabia); Prime Minister Erna Solberg (Norway); Prime Minister Shinzo Abe (Japan); President Joko Widodo (Indonesia); Prime Minister Justin Trudeau (Canada); Bill Gates; President Barack Obama (United States); President François Hollande (France); Prime Minister Narendra Modi (India); President Dilma Rousseff (Brazil); President Michelle Bachelet (Chile); Prime Minister Lars Løkke Rasmussen (Denmark); Prime Minister Matteo Renzi (Italy); President Enrique Peña Nieto (Mexico); Prime Minister David Cameron (United Kingdom); Sultan Al Jaber, Minister of State and Special Envoy for Energy and Climate Change (United Arab Emirates). Photo: Ian Langsdon/AFP via Getty Images.

Chapter 1: Why Zero?

21   Figure: Three lines you should know: Coupled Model Intercomparison Project (CMIP5) global mean temperature anomalies computed by the Royal Netherlands Meteorological Institute (KNMI) Climate Explorer. Temperature change measured in degrees Celsius.

24   Figure: Carbon emissions are on the rise: Data for average temperature change measured in degrees Celsius, relative to the 1951–1980 average is from Berkeley Earth, berkeleyearth.org; data for CO2 measured in metric tons is from Global Carbon Budget 2019 by Le Quéré, Andrew et al., which is licensed under CC BY 4.0 (https://www.creativecommons.org/licenses/by/4.0) and available at https://essd.copernicus.org/articles/11/1783/2019/.

27   Photo: AFP via Getty Images.

27   One study estimated: Solomon M. Hsiang and Amir S. Jina, "Geography, Depreciation, and Growth," *American Economic Review,* May 2015.

28   According to the U.S. government: Donald Wuebbles, David Fahey, and Kathleen Hibbard, *National Climate Assessment 4: Climate Change Impacts in the United States* (U.S. Global Change Research Program, 2017).

29   According to research cited: R. Warren et al., "The Projected Effect on Insects, Vertebrates, and Plants of Limiting Global Warming to 1.5°C Rather than 2°C," *Science,* May 18, 2018.

29   corn is especially sensitive: World of Corn website, published by the National Corn Growers Association, worldofcorn.com.

29   In Iowa alone: Iowa Corn Promotion Board website, www.iowacorn.org.

33   That drought was made three times: Colin P. Kelley et al., "Climate Change in the Fertile Crescent and Implications of the Recent Syrian Drought," *PNAS,* March 17, 2015.

33   One study that looked: Anouch Missirian and Wolfram Schlenker, "Asylum Applications Respond to Temperature Fluctuations," *Science,* Dec. 22, 2017.

Chapter 2: This Will Be Hard

38   Photo: dem10/E+ via Getty Images and lessydoang/RooM via Getty Images.

39   Here's the math: U.S. Energy Information Administration, www.eia.gov.

41   Figure: Where the emissions are: Greenhouse gases measured in metric

tons of carbon dioxide equivalents (CO2e) from Rhodium Group. This graph also uses population data from the United Nations World Population Prospects 2019, which is licensed under CC BY 3.0 IGO (https://creativecommons.org/licenses/by/3.0/igo/) and available at https://population.un.org/wpp/Download/Standard/Population/.

42   Photo: Paul Seibert.

43   Photo: ©Bill & Melinda Gates Foundation/Prashant Panjiar.

43   There are parts of Asia: Vaclav Smil, *Energy Myths and Realities* (Washington, D.C.: AEI Press, 2010), 136–37.

43   And consider how long it took: Ibid., 138.

44   Figure: It takes a really long time: Modern renewables include wind, solar, and modern biofuels. Source: Vaclav Smil, *Energy Transitions* (2018)

44   Natural gas followed: Ibid.

45   Some scientists have argued: Xiaochun Zhang, Nathan P. Myhrvold, and Ken Caldeira, "Key Factors for Assessing Climate Benefits of Natural Gas Versus Coal Electricity Generation," *Environmental Research Letters,* Nov. 26, 2014, iopscience.iop.org.

49   about 300 million tons: Rhodium Group analysis.

Chapter 3: Five Questions to Ask in Every Climate Conversation

57   Table: How much power does it take?: The figures show average power consumption. Peak demand will be higher; for example, in 2019, the peak U.S. demand was 704 gigawatts. See the U.S. Energy Administration website (www.eia.gov) for more information.

64   In the United States: Taking Stock 2020: The COVID-19 Edition, Rhodium Group, https://rhg.com.

Chapter 4: How We Plug In

67   Photo: Courtesy Gates family.

68   Figure: 860 million people don't have reliable access to electricity: Based on IEA data from IEA (2020), SDG7: Data and Projections, IEA 2020, www.iea.org/statistics. All rights reserved; as modified by Gates Ventures, LLC.

69   When you cover land: Nathan P. Myhrvold and Ken Caldeira, "Greenhouse Gases, Climate Change, and the Transition from Coal to Low-Carbon Electricity," *Environmental Research Letters,* Feb. 16, 2012, iopscience.iop.org.

70   Figure: Getting all the world's electricity: Renewables sector includes wind, solar, geothermal, and modern biofuels. Source: bp Statistical Review of World Energy 2021, https://www.bp.com.

70   One study found: Vaclav Smil, *Energy and Civilization* (Cambridge, Mass.: MIT Press, 2017), 406.

71   Photo: Universal Images Group via Getty Images

71   In all, these tax expenditures: U.S. Department of Energy Office of Scientific and Technical Information, "Analysis of Federal Incentives Used to Stimulate Energy Production: An Executive Summary," Feb. 1980, www.osti.gov. Calculation adjusts subsidies for coal and natural gas to 2019 dollars.

71   Most countries take various steps: Wataru Matsumura and Zakia Adam, "Fossil Fuel Consumption Subsidies Bounced Back Strongly in 2018," IEA commentary, June 13, 2019.

73   Europe is similarly well situated: Data derived from Eurelectric, "Decarbonisation Pathways," May 2018, cdn.eurelectric.org.

78   But Germany produced: Fraunhofer ISE, www.energy-charts.de.

78   it ends up transmitting: Zeke Turner, "In Central Europe, Germany's Renewable Revolution Causes Friction," *Wall Street Journal,* Feb. 16, 2017.

85   Figure: How much stuff does it take: Weight of materials, measured in metric tons, per terawatt-hour of electricity generated. "Solar PV" refers to solar photovoltaic panels, which convert light from the sun into electricity. Source: U.S. Department of Energy, *Quadrennial Technology Review: An Assessment of Energy Technologies and Research Opportunities* (2015), https://www.energy.gov.

87   Figure: Is nuclear power dangerous? This graph uses data from Deaths per TWh by Markandya & Wilkinson; Sovacool et al., which is licensed under CC BY 4.0 (https://www.creativecommons.org/licenses/by/4.0/) and available at https://ourworldindata.org/grapher/death-rates-from-energy-production-per-twh.

90   The United States has considerable offshore wind: U.S. Department of Energy, "Computing America's Offshore Wind Energy Potential," Sept. 9, 2016, www.energy.gov.

91   In his fantastic 2009 book: David J. C. MacKay, *Sustainable Energy— Without the Hot Air* (Cambridge, U.K.: UIT Cambridge, 2009), 98, 109.

95   And in all likelihood: Consensus Study Report, "Negative Emissions Technologies and Reliable Sequestration: A Research Agenda," National Academies of Science, Engineering, and Medicine, 2019.

Chapter 5: How We Make Things

98   Each weighs thousands of tons: Washington State Department of Transportation, www.wsdot.wa.gov.

99   Photo: WSDOT.

99   The next time you see: "Statue Statistics," Statue of Liberty National Monument, New York, National Park Service, www.nps.gov.

99   Thomas Edison tried to create: Vaclav Smil, *Making the Modern World* (Chichester, U.K.: Wiley, 2014), 36.

100   Figure: China makes a lot of cement: Measured in metric tons of cement production. Source: U.S. Department of the Interior, U.S. Geological Survey, T. D. Kelly, and G. R. Matos, comps., 2014, "Historical Statistics for Mineral and Material Commodities in the United States" (2016 version): U.S. Geological Survey Data Series 140, accessed December 6, 2019; USGS Minerals Yearbooks—China (2002, 2007, 2011, 2016), https://www.usgs.gov.

101   Plastics are also what allow: American Chemistry Council, "Plastics and Polymer Composites in Light Vehicles," Aug. 2019, www.automotiveplastics.com.

101   Photo: REUTERS/Carlos Barria.

104   China is the biggest producer: U.S. Department of the Interior, U.S. Geological Survey, "Mineral Commodity Summaries 2019."

104   Between now and 2050: Freedonia Group, "Global Cement—Demand and Sales Forecasts, Market Share, Market Size, Market Leaders," May 2019, www.freedoniagroup.com.

107   Table: Green Premiums for plastics, steel, and cement: Direct emission only; excludes emissions from electricity production. Source: Rhodium Group.

Chapter 6: How We Grow Things

113   We'll also have to do something: Rhodium Group internal analysis.

114   "The battle to feed": Paul Ehrlich, *The Population Bomb* (New York: Ballantine Books, 1968).

114   In the time since: World Bank, data.worldbank.org.

114   the average household spends less: Derek Thompson, "Cheap Eats: How America Spends Money on Food," *The Atlantic,* March 8, 2013, www.theatlantic.com.

116   Figure: Most countries aren't consuming more meat: Consumption

measured in metric tons of meat, which includes beef, lamb, pork, poultry, and veal. Source: OECD (2020), OECD-FAO Agricultural Outlook, https://stats.oecd.org (accessed October 2020).

117   Around the world: Food and Agriculture Organization of the United Nations, www.fao.org.

119   "The gastronomic meal": UNESCO, "Gastronomic Meal of the French," ich.unesco.org.

119   On average, a ground-beef substitute: Online survey of U.S. retail prices in September 2020 conducted by Rhodium Group.

122   Photo: Gates Notes, LLC.

123   Figure: There's a huge gap in agriculture: Measured in thousands of kilograms (kg) of corn per hectare (ha). Source: Food and Agriculture Organization of the United Nations. FAOSTAT. OECD-FAO Agricultural Outlook 2020-2029. Latest update: November 30, 2020. Accessed: November 2020. https://stats.oecd.org/Index.aspx?datasetcode=HIGH _AGLINK_2020#.

126   According to the World Bank: World Bank Development Indicators, databank.worldbank.org.

126   One study by the World Resources Institute: Janet Ranganathan et al., "Shifting Diets for a Sustainable Food Future," World Resources Institute, www.wri.org.

127   It's one of the main reasons: World Resources Institute, "Forests and Landscapes in Indonesia," www.wri.org.

Chapter 7: How We Get Around

132   Organization for Economic Cooperation and Development predicts: https://www.oecd-ilibrary.org/.

133   Figure: COVID-19 is slowing: Historical emissions provided by Rhodium Group. Projected emissions based on IEA data from IEA (2020), World Energy Outlook, IEA 2020, www.iea.org/statistics. All rights reserved; as modified by Gates Ventures, LLC.

134   Figure: Cars aren't the only culprit. Source: This chart uses data from Beyond road vehicles: Survey of zero-emission technology options across the transport sector by Hall, Pavlenko, and Lutsey, which is licensed under CC BY-SA 3.0 (https://www.creativecommons.org/licenses/by-sa /3.0/) and available at https://theicct.org/sites/default/files/publications /Beyond_Road_ZEV_Working_Paper_20180718.pdf.

135   There are about a billion cars: International Organization of Motor Vehicle Manufacturers (OICA), www.oica.net.

135   In 2018 alone: This assumes a gross addition of 69 million cars per year per OICA and retirements of about 45 million, with a vehicle life span of 13 years.

136   Figure: Chevy versus Chevy: Specifications for Chevrolet Malibu and Bolt EV are for model year 2022. Source: https://www.chevrolet.com. Illustrations ©izmocars—All rights reserved.

136   When you account for: Price per mile assumes buyer pays an average purchase price for the car, owns it for seven years, and drives an average of 12,000 miles per year. Source: Rhodium Group.

139   Table: Green Premium to replace gasoline with advanced biofuels: Rhodium Group, Evolved Energy Research, IRENA, and Agora Energiewende. Retail price is the average in the United States from 2015 to 2018. Zero-carbon option is current estimated price.

140   Table: Green Premiums to replace gasoline with zero-carbon alternatives: Rhodium Group, Evolved Energy Research, IRENA, and Agora Energiewende. Retail price is the average in the United States from 2015 to 2018. Zero-carbon option is current estimated price.

140   a typical household spends: U.S. Energy Information Administration, www.eia.gov.

140   The city of Shenzhen: Michael J. Coren, "Buses with Batteries," *Quartz*, Jan. 2, 2018, www.qz.com.

141   Photo: Bloomberg via Getty Images.

141   According to a 2017 study: Shashank Sripad and Venkatasubramanian Viswanathan, "Performance Metrics Required of Next-Generation Batteries to Make a Practical Electric Semi Truck," *ACS Energy Letters*, June 27, 2017, pubs.acs.org.

142   Table: Green Premiums to replace diesel: Rhodium Group, Evolved Energy Research, IRENA, and Agora Energiewende. Retail price is the average in the United States from 2015 to 2018. Zero-carbon option is current estimated price.

143   Meanwhile, a mid-capacity Boeing 787: Boeing, www.boeing.com.

143   Table: Green Premiums to replace jet fuel: Rhodium Group, Evolved Energy Research, IRENA, and Agora Energiewende. Retail price is the average in the United States from 2015 to 2018. Zero-carbon option is current estimated price.

143   The same goes for cargo ships: Kyree Leary, "China Has Launched the World's First All-Electric Cargo Ship," Futurism, Dec. 5, 2017, futurism .com; "MSC Receives World's Largest Container Ship MSC Gulsun from SHI," Ship Technology, July 9, 2019, www.ship-technology.com.

144   Table: Green Premiums to replace bunker fuel: Rhodium Group,

Evolved Energy Research, IRENA, and Agora Energiewende. Retail price is the average in the United States from 2015 to 2018. Zero-carbon option is current estimated price.

144   Table: Green Premiums to replace current fuels: Rhodium Group, Evolved Energy Research, IRENA, and Agora Energiewende. Retail price is the average in the United States from 2015 to 2018. Zero-carbon option is current estimated price.

146   In 2019, we bought more than 5 million cars: S&P Global Market Intelligence, https://www.spglobal.com/marketintelligence/en/.

Chapter 8: How We Keep Cool and Stay Warm

148   Humans have been trying: A. A'zami, "Badgir in Traditional Iranian Architecture," Passive and Low Energy Cooling for the Built Environment conference, Santorini, Greece, May 2005.

148   But the first known machine: U.S. Department of Energy, "History of Air Conditioning," www.energy.gov. Also "The Invention of Air Conditioning," *Panama City Living,* March 13, 2014, www.panamacityliving .com.

149   Barely more than a century: International Energy Agency, "The Future of Cooling," www.iea.org.

150   Worldwide, there are 1.6 billion: International Energy Agency, www.iea .org.

150   Figure: A/C is on the way: Based on IEA data from IEA (2018), The Future of Cooling, IEA (2018), www.iea.org/statistics. All rights reserved; as modified by Gates Ventures, LLC.

150   Worldwide, sales rose 15 percent: Ibid.

151   in the United States: U.S. Environmental Protection Agency, www.epa .gov.

154   Table: Green Premium for installing an air-sourced heat pump: Rhodium Group. This table shows the net present value of an air-sourced heat pump versus a natural gas heater and an electric A/C in a new house. Costs are calculated using a 7 percent discount rate and current prices for electricity and natural gas as of summer 2019 and a 15-year life span for the heat pump.

154   If heat pumps are such a great deal: U.S. Energy Information Administration, www.eia.gov.

156   Table: Green Premiums to replace current heating fuels: Rhodium Group, Evolved Energy Research, IRENA, and Agora Energiewende.

Retail price is the average in the United States from 2015 to 2018. Zero-carbon option is current estimated price.

156   If their home is heated: Ibid.

157   An extreme example: Bullitt Center, www.bullittcenter.org.

158   Photo: Nic Lehoux.

Chapter 9: Adapting to a Warmer World

161   Photo: ©Bill & Melinda Gates Foundation/Frederic Courbet.

162   Worldwide, there are 500 million: Max Roser, Our World in Data website, ourworldindata.org.

163   The typical Kenyan: World Bank, www.data.worldbank.org.

164   The world knows a lot: GAVI, www.gavi.org.

167   In fact, doubling CGIAR's funding: Global Commission on Adaptation, *Adapt Now: A Global Call for Leadership on Climate Resilience*, World Resources Institute, Sept. 2019, gca.org.

168   Photo: From the photo collection of the International Rice Research Institute (IRRI), Los Banos, Laguna, Philippines.

169   The payoff could be dramatic: Food and Agriculture Organization of the United Nations, *State of Food and Agriculture: Women in Agriculture, 2010–2011,* www.fao.org.

173   Photo: Mazur Travel via Shutterstock.

175   Extreme poverty has plummeted: World Bank, "Decline of Global Extreme Poverty Continues but Has Slowed," www.worldbank.org.

Chapter 10: Why Government Policies Matter

180   Photo: Mirrorpix via Getty Images.

182   As a result of these: U.S. Energy Information Administration, www.eia.gov.

192   Germany gave the market a boost: International Energy Agency.

192   Then, in 2011, the United States: U.S. Energy Department, "Renewable Energy and Efficient Energy Loan Guarantees," www.energy.gov.

193   Photo: Sirio Magnabosco/EyeEm via Getty Images.

Chapter 11: A Plan for Getting to Zero

201   The project took 13 years: Human Genome Project Information Archive, "Potential Benefits of HGP Research," web.ornl.gov.

201   An independent study: Simon Tripp and Martin Grueber, "Economic Impact of the Human Genome Project," Battelle Memorial Institute, www.battelle.org.

Chapter 12: What Each of Us Can Do

226   "When we have a fact-based worldview": Hans Rosling, *Factfulness: Ten Reasons We're Wrong About the World—and Why Things Are Better than You Think,* with Ola Rosling and Anna Rosling Rönnlund (New York: Flatiron Books, 2018), 255.

Afterword: Climate Change and COVID-19

228   Black people and Latinx people: "Race, Ethnicity, and Age Trends in Persons Who Died from COVID-19—United States, May–August 2020," U.S. Centers for Disease Control https://www.cdc.gov.

228   Among recipients of Medicare: "Preliminary Medicare COVID-19 Data Snapshot," Centers for Medicare and Medicaid Services, https://www.cms.gov.

229   A study by the Institute for Health Metrics and Evaluation: "Goalkeepers Report 2020," https://www.gatesfoundation.org.

229   In 2018, the U.S. government's investment: "Impacts of Federal R&D Investment on the U.S. Economy," Breakthrough Energy, https://www.breakthroughenergy.org.

Page numbers in *italics* refer to illustrations.